Fundamentals of Ramsey Theory

Discrete Mathematics and Its Applications
Series editors:
Miklos Bona, Donald L. Kreher, Douglas B. West

Algorithmics of Nonuniformity
Tools and Paradigms
Micha Hofri, Hosam Mahmoud

Handbook of Geometric Constraint Systems Principles
Edited by Meera Sitharam, Audrey St. John, Jessica Sidman

Introduction to Chemical Graph Theory
Stephan Wagner, Hua Wang

Extremal Finite Set Theory
Daniel Gerbner, Balazs Patkos

The Mathematics of Chip-Firing
Caroline J. Klivans

Computational Complexity of Counting and Sampling
Istvan Miklos

Volumetric Discrete Geometry
Karoly Bezdek, Zsolt Langi

The Art of Proving Binomial Identities
Michael Z. Spivey

Combinatorics and Number Theory of Counting Sequences
Istvan Mezo

Applied Mathematical Modeling
A Multidisciplinary Approach
Douglas R. Shier, K.T. Wallenius

Analytic Combinatorics
A Multidimensional Approach
Marni Mishna

50 years of Combinatorics, Graph Theory, and Computing
Edited By Fan Chung, Ron Graham, Frederick Hoffman, Ronald C. Mullin, Leslie Hogben, Douglas B. West

Fundamentals of Ramsey Theory
Aaron Robertson

https://www.routledge.com/Discrete-Mathematics-and-Its-Applications/book-series/CHDISMTHAPP.

Fundamentals of Ramsey Theory

Aaron Robertson
Colgate University

CRC Press
Taylor & Francis Group
Boca Raton London New York

CRC Press is an imprint of the
Taylor & Francis Group, an **informa** business
A CHAPMAN & HALL BOOK

First edition published 2021
by CRC Press
6000 Broken Sound Parkway NW, Suite 300, Boca Raton, FL 33487-2742

and by CRC Press
2 Park Square, Milton Park, Abingdon, Oxon, OX14 4RN

Library of Congress Cataloging-in-Publication Data

Names: Robertson, Aaron, 1971- author.
Title: Fundamentals of Ramsey theory / Aaron Robertson, Colgate University.
Description: First edition. | Boca Raton : Chapman & Hall, CRC Press, 2021.
| Series: Discrete mathematics and its applications | Includes
bibliographical references and index.
Identifiers: LCCN 2021000970 (print) | LCCN 2021000971 (ebook) | ISBN
9781138364332 (hardback) | ISBN 9780429431418 (ebook)
Subjects: LCSH: Ramsey theory.
Classification: LCC QA166 .R57 2021 (print) | LCC QA166 (ebook) | DDC
511.5--dc23
LC record available at https://lccn.loc.gov/2021000970
LC ebook record available at https://lccn.loc.gov/2021000971

ISBN: 9781138364332 (hbk)
ISBN: 9780429431418 (ebk)

To my advisor, Doron Zeilberger

Contents

Preface

Ramsey theory has fascinated me since being introduced to it in graduate school. My first encounter was through the book by Graham, Rothschild, and Spencer [84]. Since their book was published, many books on Ramsey theory have been produced. So, do we really need another book? Probably not; however, my goal in writing this book is to help give an overview of Ramsey theory from several points of view, adding intuition and detailed proofs as we go, while being, hopefully, a bit gentler than most of the other books on Ramsey theory as a whole.

I would certainly be remiss if I did not acknowledge the profound impact of those who have propelled my interest in Ramsey theory, both implicitly and explicitly: Paul Erdős, Ron Graham (who unfortunately passed away while this book was being written), Joel Spencer, Bruce Landman, and Tom Brown. A special debt of gratitude goes to my advisor, Doron Zeilberger, to whom this book is dedicated. For whatever reason, he gave me a problem in Ramsey theory, which was not the typical type of problem he would present to his students. The problem was a Ron Graham \$100 prize problem, which was simultaneously, and independently, solved by Tomasz Schoen. Upon publication, I received a check from Ron. In hindsight, I should not have cashed that check, but I suspect I was probably hungry as a poor graduate student.

While writing this book, I assumed a reader comfortable with real and complex analysis, linear and abstract algebra, topology, probability, statistics, number theory, and combinatorics at the undergraduate level. In general, a solid undergraduate training in mathematics should suffice.

This book covers all of the main results in Ramsey theory that are found in other books on the subject along with results that have not appeared in a book before. The book begins by presenting some basic results in combinatorics, analysis, Fourier analysis, probability, abstract algebra, and topology that will be used throughout the book. Chapter 2 presents results on the integers, in particular, van der Waerden's Theorem, Rado's Theorem, and the Hales-Jewett Theorem. In Chapter 3, we finally get to the theorem for which Ramsey theory is named. Some results concerning the colorings of space appear in Chapter 4. Chapter 5 turns away from the combinatorial flavor that dominates the preceding chapters and presents previously encountered results through different lenses. The probabilistic method as it applies to Ramsey-type problems, including the powerful Lovász Local Lemma, is encountered in Chapter 6. The book ends with a few applications of Ramsey theory.

Aaron Robertson

Symbols

Symbol Description

$\lvert \cdot \rvert$	cardinality of a set
$(\cdot, \cdot)$	greatest common divisor
$\sqcup$	disjoint union
$\boxtimes$	strong product of graphs
$\odot$	dot product
$\oplus$	ultrafilter addition
$\alpha(\cdot)$	independence number of a graph
$\mathrm{Aut}(\cdot)$	automorphism group of a graph
$\beta\mathbb{Z}^+$	set of ultrafilters of $\mathbb{Z}^+$
$\mathbb{C}$	complex numbers
$\chi(G)$	chromatic number of graph G
$\chi(t)$	additive character $e^{\frac{2\pi i t}{n}}$
$\mathbb{E}(\cdot)$	expectation
$\mathbb{F}_p$	finite field of size p
$\widehat{f}(t)$	discrete Fourier transform
G^*	dual graph of G
$HJ(k;r)$	Hales-Jewett number
K_n	complete graph on n vertices
$\omega(\cdot)$	clique number of a graph
$\wp(\cdot)$	power set
$\mathbb{P}(\cdot)$	probability
$\mathbb{Q}$	rational numbers
$\mathbb{R}$	real numbers
$r_k(n)$	maximum size of a subset of $[1,n]$ with no k-term arithmetic progression
$R(k;r)$	Ramsey number
$s(r)$	Schur number
$\mathrm{Sym}(\cdot)$	symmetric group
$w(k;r)$	van der Waerden number
$\mathbb{Z}$	integers
$\mathbb{Z}^+$	positive integers
$\mathbb{Z}^m$	$\mathbb{Z} \times \mathbb{Z} \times \cdots \times \mathbb{Z}$ (m copies)
$\mathbb{Z}_p$	integers modulo p
$\mathbb{Z}_p^*$	group of $\mathbb{Z}_p \setminus \{0\}$ under multiplication

1

Introduction

There is a light that never goes out.

–Morrissey

Unbreakable, inevitable, assured, inescapable: choose whichever synonym you prefer. Ramsey theory is the study of properties that must occur for sufficiently large structures. We can try to break the structure through partitioning, but a Ramsey property will persist.

1.1 What is Ramsey Theory?

Consider positive integer solutions to $x + y = z$. Of course, $1 + 1 = 2$ and $2 + 3 = 5$ work if we allow all positive integers. So, let's try to break this by splitting the positive integers into two parts. In Ramsey theory, we typically use colors to describe the partitions, so we will have, say, red integers and blue integers.

Must we still have a solution to $x + y = z$ if we now require the integers to be in the same partition, i.e., the same color? Let's see if we can avoid the property of one part of the partition having a solution to $x + y = z$. First, 1 and 2 must be different colors (since $1 + 1 = 2$) and, consequently, 4 must be the same color as 1 (since $2 + 2 = 4$). Let's say that 1 and 4 are red and 2 is blue. Since $1 + 4 = 5$, we see that 5 must also be blue, and, consequently, 3 must be red (since $2 + 3 = 5$). But now $1, 3$, and 4 are all red, so the Ramsey property persists.

Ramsey properties also exist on graphs. For example, if we take $n \geq 3$ vertices and connect every pair of vertices with an edge, we clearly have a triangle with all edges in the same partition. Can we partition the edges in such a way so that we no longer have a triangle with all edges in the same partition? The answer is no, provided we have at least 6 vertices. To see this, isolate one vertex, say X. Using the colors red and blue, since X is connected to at least 5 other vertices, we see that one of the colors must occur on at least 3 of the edges. Let X be connected to each of vertices A, B, and C with a blue edge. If any edge between any two of A, B, and C is blue, then we have a blue triangle. Otherwise, all edges among A, B, and C are red and we again have a monochromatic (red) triangle. So, as long as we have enough vertices (here, we have shown that 6 vertices suffice), we cannot avoid a monochromatic triangle. It is easy to show that 6 vertices is also necessary by considering only 5 vertices $V_0, V_1, \ldots, V_4$ and letting the edge between V_i and V_j be red if $j \equiv i+1 \pmod 5$, and blue otherwise.

The above examples are the easiest cases of two of the most well-known theorems in Ramsey theory: Schur's Theorem ($x + y = z$) and Ramsey's Theorem (graphs). If these intrigue you, then you will find other compelling results in the coming chapters.

1.2 Notations and Conventions

We will use the following notations and conventions. Other symbols that are used appear on page xiii.

Since Ramsey theory deals with colorings, formally, for $r \in \mathbb{Z}^+$, an *r-coloring* of the elements of a set S is a mapping $\chi : S \to T$, where $|T| = r$ and, typically, $T = \{0, 1, \ldots, r-1\}$.

Unless otherwise stated, the intervals we use are integer intervals. Hence, we rely on the notation

$$[1, n] = \{1, 2, \ldots, n\}.$$

We may refer to this interval by $\mathbb{Z}_n$ if we are doing arithmetic modulo n. Similarly, we will assume all arithmetic progressions are of integers unless otherwise stated.

As hinted at by the definition of an r-coloring, we will be dealing with sets often. For any set S, we use $|S|$ to denote the cardinality of S and we use $\wp(S)$ to denote the power set of S.

In our asymptotic analysis, we will use the notations $O(n), o(1)$, and $\ll$. We remind the reader of these next.

Definition 1.1 (Big-O, Little-o, and $\ll$). We say that $f(n) = O(n)$ if there exists a constant $c > 0$ such that $|f(n)| \leq cn$ for all sufficiently large n. We say that $f(n) = o(1)$ if $\lim_{n\to\infty} f(n) = 0$. We say $f(n) \ll g(n)$ if $\lim_{n\to\infty} \frac{f(n)}{g(n)} = 0$.

We will also be using logarithms often. For our purposes, log is the base 2 logarithm and ln is the natural logarithm.

When doing probability, the probability of event E is denoted $\mathbb{P}(E)$; the expectation of random variable X is denoted $\mathbb{E}(X)$.

1.3 Prerequisites

At various spots in this book, results from areas as disparate as Fourier analysis, abstract algebra, and statistics will be used. Below, we list some results that may be used without reference and with which we assume the reader is knowledgeable. The preliminaries given here are not intended to be an exhaustive list; rather, the goal is to give the reader an idea of the mathematical background assumed.

1.3.1 Combinatorics

When doing asymptotic analysis, we will rely on two results. The first is Stirling's formula, formulated as

$$n! \approx \sqrt{2\pi n}\left(\frac{n}{e}\right)^n.$$

The second is

$$\ln(1+x) \approx x$$

for x small. Some consequences of these appear in the Exercises section of this chapter, and those consequences will be used without reference.

Many results in Ramsey theory have their origins with the pigeonhole principle. This principle is obvious in statement, but not necessarily so in application. For reference, here is the pigeonhole principle.

Theorem 1.2 (Pigeonhole Principle). *Let $k, r \in \mathbb{Z}^+$. If $n \geq kr+1$ elements are partitioned into r parts, then one of those parts must contain at least $k+1$ elements.*

The last combinatorial concept we remind the reader of is the Principle of Inclusion-Exclusion. It is a counting principle but can also be stated in terms of probabilities (see Equation 6.1).

Theorem 1.3 (Principle of Inclusion-Exclusion). *Let $S_1, S_2, \ldots, S_m$ be finite sets. Then*

$$\begin{aligned}\left|\bigcup_{i=1}^m S_i\right| &= \sum_{i=1}^m |S_i| - \sum_{1\leq i<j\leq m} |S_i \cap S_j| + \sum_{1\leq i<j<k\leq m} |S_i \cap S_j \cap S_k| - \cdots + (-1)^{m+1}\left|\bigcap_{i=1}^m S_i\right| \\ &= \sum_{i=1}^m \sum_{\substack{I\subseteq[1,m]\\ |I|=i}} (-1)^{i+1}\left|\bigcap_{j\in I} S_j\right|.\end{aligned}$$

In Ramsey theory it is usually not feasible to calculate all terms in the inclusion-exclusion formula. Hence, we will use the following Bonferroni inequalities.

Theorem 1.4 (Bonferroni inequalities). *Let $S_1, S_2, \ldots, S_m$ be finite sets. Then*

$$\left|\bigcup_{i=1}^{m} S_i\right| \leq \sum_{i=1}^{m} |S_i|$$

and

$$\left|\bigcup_{i=1}^{m} S_i\right| \geq \sum_{i=1}^{m} |S_i| - \sum_{1\leq i<j\leq m} |S_i \cap S_j|.$$

1.3.2 Analysis

We start by reminding the reader of the definition of a metric.

Definition 1.5 (Metric, Triangle inequality). Let X be a set. We say that $d : X \times X \to \mathbb{R}$ is a *metric* on X if the following are satisfied for all $x, y, z \in X$:

(i) $d(x, y) \geq 0$;

(ii) $d(x, y) = 0$ if and only if $x = y$;

(iii) $d(x, y) = d(y, x)$;

(iv) $d(x, y) \leq d(x, z) + d(z, y)$ (this is referred to as the *triangle inequality*).

We will have the need to appeal to Fekete's Lemma, which is quite useful for many combinatorial functions, not just in Ramsey theory.

Lemma 1.6 (Fekete's Lemma). *For any sequence of real numbers $\{s_i\}_{i=1}^{\infty}$, if either (i) $s_{i+j} \leq s_i + s_j$ for all $i, j \in \mathbb{Z}^+$ or (ii) $s_{i+j} \geq s_i + s_j$ for all $i, j \in \mathbb{Z}^+$, then*

$$\lim_{n\to\infty} \frac{s_n}{n}$$

exists and equals $\inf_n \frac{s_n}{n}$ if (i) is satisfied; it equals $\sup_n \frac{s_n}{n}$ if (ii) is satisfied.

An easy corollary of Fekete's Lemma is also useful (and is often referred to as Fekete's Lemma, too).

Corollary 1.7. *For any sequence of real numbers $\{s_i\}_{i=1}^{\infty}$, if either (i) $s_{i+j} \leq s_i \cdot s_j$ for all $i, j \in \mathbb{Z}^+$ or (ii) $s_{i+j} \geq s_i \cdot s_j$ for all $i, j \in \mathbb{Z}^+$, then*

$$\lim_{n\to\infty} (s_n)^{\frac{1}{n}}$$

exists and equals $\inf_n (s_n)^{\frac{1}{n}}$ if (i) is satisfied; it equals $\sup_n (s_n)^{\frac{1}{n}}$ if (ii) is satisfied.

Remark. A sequence satisfying condition (i) of Lemma 1.6 is called subadditive (condition (ii) is referred to as superadditive); a sequence satisfying condition (i) of Corollary 1.7 is called sub-multiplicative (condition (ii) is referred to as super-multiplicative).

Fourier analysis is not a standard topic in many undergraduate analysis courses. Hence, this part of the Prerequisites section will be a little more detailed.

As notation, we will always remind the reader in sections using Fourier analysis that we reserve i for $i = \sqrt{-1}$. Also, for $c \in \mathbb{C}$, we denote by $\overline{c}$ the complex conjugate of c; that is, if $c = a + ib$, then $\overline{c} = a - ib$ and, in polar form, if $c = re^{i\theta}$ then $\overline{c} = re^{-i\theta}$.

Many results we present that involve analysis revolve around the concept of upper density, which is defined as follows.

Definition 1.8 (Upper density). Let $S \subseteq \mathbb{Z}^+$. The *upper density of* S, denoted $\overline{d}(S)$, is defined as

$$\overline{d}(S) = \limsup_{n\to\infty} \frac{|S \cap [1,n]|}{n}.$$

We will also use the concept of *lower density* in the proof of Theorem 2.67, where lim sup is replaced by lim inf.

With Fourier analysis, we will be working on finite intervals $[1, n]$ and will need to consider the characters of $(\mathbb{Z}_n, +)$, which are the nth roots of unity:

$$\chi(t) = e^{\frac{2\pi i t}{n}}.$$

We will refer to these as *additive characters* to emphasize that they are homomorphisms from the additive group $(\mathbb{Z}_n, +)$ to the multiplicative group $\mathbb{C} \setminus \{0\}$.

We know that $|\chi(t)| = 1$ for any $t \in \mathbb{R}$, while $\chi(t) = 1$ if and only if $t = 0$. Furthermore, for any integer j, we have

$$\frac{1}{n}\sum_{k=0}^{n-1} \chi(kj) = \begin{cases} 0 & j \neq 0; \\ 1 & j = 0. \end{cases}$$

Letting $j = 1$, we can consider

$$\lim_{n\to\infty} \frac{1}{n}\sum_{k=0}^{n-1} \chi(k) = \lim_{n\to\infty} \sum_{k=0}^{n-1} e^{2\pi i \frac{k}{n}} \cdot \frac{1}{n} = 0$$

and, noting that this is a Riemann sum, deduce that

$$\int_0^1 e^{2\pi i x}\, dx = \begin{cases} 0 & x \neq 0; \\ 1 & x = 0. \end{cases}$$

We now define the discrete Fourier transform of a function $f : \mathbb{Z}_n \to \mathbb{C}$ by

$$\begin{aligned}\widehat{f}(t) &= \frac{1}{n}\sum_{k=0}^{n-1} f(k)\overline{\chi(kt)} \\ &= \frac{1}{n}\sum_{k=0}^{n-1} f(k)\chi(-kt),\end{aligned}$$

where $\widehat{f} : \mathbb{Z}_n \to \mathbb{C}$.

The first expression informs us that the Fourier transform is similar to the standard inner product for complex-valued functions over $\mathbb{Z}_n$: $\langle f, g\rangle = \sum_{k=0}^{n-1} f(k)\overline{g(k)}$. Hence, the Fourier transform inherits many inner product properties. We further note that this can also be viewed, in the discrete setting, as a change of basis.

We will use the following results about the discrete Fourier transform when dealing with density results.

Theorem 1.9 (Basic Fourier Results). *Let $f, g : \mathbb{Z}_n \to \mathbb{C}$. Then the following hold:*

(i) Recovery:

$$f(j) = \sum_{k=0}^{n-1} \widehat{f}(k)\chi(kj);$$

(ii) Estimate:

$$|\widehat{f}(j)| \le \frac{1}{n}\sum_{k=0}^{n-1} |f(k)| \quad \textit{for any} \quad j \in \mathbb{Z}_n;$$

(iii) Plancherel:

$$\sum_{k=0}^{n-1} |\widehat{f}(k)|^2 = \frac{1}{n}\sum_{k=0}^{n-1} |f(k)|^2;$$

(iv) Convolution: For $j \in \mathbb{Z}_n$, defining

$$(f * g)(j) = \sum_{k=0}^{n-1} f(k)g(j-k)$$

to be the convolution of f and g, we have

$$\widehat{(f * g)}(k) = n\widehat{f}(k)\widehat{g}(k).$$

We will be applying Theorem 1.9 to sets $A \subseteq [1, n]$ by defining the indicator function $A(j) = 1$ if $j \in A$ and $A(j) = 0$ if $j \notin A$.

1.3.3 Probability

The probability we use is basic. Recall that if E and F are independent events, then

$$\mathbb{P}(E \cap F) = \mathbb{P}(E) \cdot \mathbb{P}(F),$$

but that for general events, we have

$$\mathbb{P}(E \cap F) = \mathbb{P}(E) \cdot \mathbb{P}(F|E).$$

If E and F are mutually exclusive, i.e., $E \cap F = \emptyset$, then

$$\mathbb{P}(E \sqcup F) = \mathbb{P}(E) + \mathbb{P}(F),$$

while for general events, we have

$$\mathbb{P}(E \cup F) = \mathbb{P}(E) + \mathbb{P}(F) - \mathbb{P}(E \cap F).$$

We will also use expectation of a random variable. If X is a random variable taking on possible values $x_1, x_2, \ldots$, then

$$\mathbb{E}(X) = \sum_i x_i \mathbb{P}(X = x_i).$$

We will often use indicator random variables (i.e., Bernoulli random variable, which take on values of 0 and 1 only). For any indicator random variable X, we have

$$\mathbb{E}(X) = \mathbb{P}(X = 1),$$

since $\mathbb{E}(X) = 0 \cdot \mathbb{P}(X = 0) + 1 \cdot \mathbb{P}(X = 1)$.

We will almost exclusively be dealing with finite sample spaces that have equally likely outcomes so that when we randomly choose an element from a sample space with n elements, the probability of choosing that element is $\frac{1}{n}$.

1.3.4 Algebra

We will use some linear algebra but will remind the reader of the relevant facts as needed.

Our main reminder regarding abstract algebra is for what occurs in Sections 3.3.3 and 7.1, where we use the coset decompositions of groups. For completeness, let H be a subgroup of group G. Then a (left) coset of H in G has form

$$aH = \{ah : h \in H\}$$

where $a \in G$. As far as cosets are concerned, we will only be using left cosets (and, mostly, our groups will be Abelian so that the left/right distinction is immaterial). By Lagrange's Theorem, we know that every coset of H has the same number of elements, namely $|H|$, and that no two cosets of H have non-empty intersection. It follows that the number of cosets of H in G is

$$|G : H| = \frac{|G|}{|H|}.$$

We will also be using group actions; that is, if G is a group and S is a set, we use $* : G \times S \to S$ (akin to a binary operation). Applying group actions, we will be using the concepts of orbits and stabilizers, defined next.

Definition 1.10 (Orbit)**.** Let $*$ be a group action on set S by group G. For $s \in S$, the *orbit* of s is

$$\mathcal{O}_s = \{t \in S : g * s = t \text{ for some } g \in G\}.$$

Definition 1.11 (Stabilizer)**.** Let $*$ be a group action on set S by group G. For $s \in S$, the *stabilizer of* s is

$$G_s = \{g \in G : g * s = s\}.$$

In Exercise 1.17, the reader is asked to prove that G_s is a subgroup of G.

In Section 3.3.3, we will be appealing to the Orbit-Stabilizer Theorem:

Theorem 1.12 (Orbit-Stabilizer Theorem)**.** *Let G be a finite group acting on a finite set S. Then*

$$|\mathcal{O}_s| \cdot |G_s| = |G|$$

for any $s \in S$.

1.3.5 Topology

In Chapter 5 we will be using a fair bit of topology. We have attempted to create as self-contained of a chapter as possible vis-à-vis topology. We remind the reader here of some of the concepts we will be using. The main concept we will be using is that of a compact space.

Definition 1.13 (Compact)**.** Let X be a topological space. We say that X is compact if and only if every open cover of X admits a finite subcover.

In the situation when X is a Euclidean space, we have the following useful characterization of compact.

Theorem 1.14 (Heine-Borel Theorem)**.** *Let X be a subset of $\mathbb{R}^n$ (for some $n \in \mathbb{Z}^+$). Then X is compact if and only if X is closed and bounded.*

When Theorem 1.14 does not apply but the space is still a metric space, we will use Theorem 1.17, below, which relies on the next two definitions.

Definition 1.15 (Cauchy sequence, Complete space)**.** Let $\{s_i\}_{i=1}^{\infty}$ be a sequence of elements from a metric space X with metric d. We say that the sequence is a *Cauchy sequence* if for any $\epsilon > 0$ there exists $N = N(\epsilon) \in \mathbb{Z}^+$ such that $d(s_n, s_m) < \epsilon$ for all $n, m > N$. If every Cauchy sequence converges to an element in X, we say that X is *complete.*

Definition 1.16 (Totally bounded)**.** Let X be a metric space. We say that X is *totally bounded* if for any $\epsilon > 0$ there exists a finite number of subsets $A_1, A_2, \ldots, A_n \subseteq X$ with $\text{diam}(A_i) < \epsilon$ for $1 \leq i \leq n$ such that $\bigcup_{i=1}^{n} A_i = X$.

Theorem 1.17. *Let X be a metric space. If X is complete and totally bounded, then X is compact.*

We also remind the reader of the following definitions.

Definition 1.18 (Continuous)**.** Let X and Y be topological spaces. We say that $f : X \to Y$ is *continuous* if for every open set $U \subseteq Y$ we have that $f^{-1}(U)$ is an open set in X.

Definition 1.19 (Homeomorphism)**.** Let X and Y be topological spaces. Let $f : X \to Y$ satisfy the following conditions:

(i) f is a bijection;

(ii) f is continuous;

(iii) f^{-1} is continuous.

Then f is called a *homeomorphism* (or topological isomorphism).

1.3.6 Statistics

We will be concerned with hypothesis testing in Section 7.3; specifically, does an observed count distribution behave as a random distribution would (under some additional restrictions)? This is a χ^2 goodness-of-fit test where the degrees of freedom equals one less than the number of comparison categories.

We also rely on the standard test to compare two independent chi-squared statistics: the F-test.

1.3.7 Practice

To prepare the reader for some of the concepts that will be used in subsequent chapters, we strongly urge the reader to do all of the exercises at the end of this chapter, all of which involve concepts that the reader will encounter later in this book.

1.4 Compactness Principle

There is an interplay between finite and infinite Ramsey theory. While much of this book focuses on finite Ramsey theory, we can use the infinite to prove the finite. This is accomplished by the Compactness Principle, which we state below.

Compactness Principle. *Let $\mathcal{F}$ be a family of finite subsets of $\mathbb{Z}^+$. Let $k, r \in \mathbb{Z}^+$. Assume that every r-coloring of the k-element subsets of $\mathbb{Z}^+$ admits $F \in \mathcal{F}$ with the property that all k-element subsets of F have the same color. Then there exists $N \in \mathbb{Z}^+$ such that for all $n \geq N$, any r-coloring of the k-element subsets of $[1, n]$ admits $G \in \mathcal{F}$ with $G \subseteq [1, n]$ such that the collection of k-element subsets of G is monochromatic.*

This result can be used at times to bypass technical details in proofs. Since many Ramsey theory results are about having "large enough" systems, when dealing with the set of integers we often encounter statements of the form "for all $n \geq N$, property P holds." If we can show that property P holds over the positive integers, then the Compactness Principle gives us the "for all $n \geq N$" part of the statement.

The proof of the Compactness Principle is essentially Cantor's argument for proving that the set of real numbers is uncountable; the reader is referred to [129] for a proof.

A word of warning is warranted here. The Compactness Principle does not work in reverse; that is, we cannot prove the finite version and conclude that it holds for the infinite. As we will see, results on infinite sets can run counter to their finite counterparts. Just keep in mind that arbitrarily large does not mean infinite.

1.5 Set Theoretic Considerations

We will be focusing our attention on countable objects; however, Ramsey theory is also studied over uncountable sets. We can ask, for example, do similar results hold if we color $\mathbb{R}^+$ instead of $\mathbb{Z}^+$? When doing so, the Axiom of Choice (or one of its equivalents) may come into play. We will largely (but not always) stay away from this uncountable territory; see [117] for a recent treatment of Ramsey theory in the uncountable setting.

We will note here that, like the Banach-Tarski paradox, we get some strange results in the uncountable setting. Consider the set of infinite subsets of $\mathbb{Z}^+$, denoted by $\mathcal{P}$. Assume that each $P \in \mathcal{P}$ is assigned one of two colors. In order for $\mathcal{P}$ to have the Ramsey property, we would require that under every possible 2-coloring χ of the elements of $\mathcal{P}$, there exists an infinite set $S \in \mathcal{P}$ such that all infinite subsets of S have the same color under χ.

As noted by Galvin and Prikry [79], both of the following hold:

(i) assuming the Axiom of Choice is true (so that there exist subsets of $\mathbb{R}$ that are non-measurable), Erdős and Rado [68] have provided a 2-coloring of the elements of $\mathcal{P}$ such that no infinite set has all infinite subsets the same color;

(ii) assuming the Axiom of Choice is false and that all subsets of $\mathbb{R}$ are measurable, Mathias [142] has shown that under any 2-coloring of the elements of $\mathcal{P}$ there exists an infinite set that has all infinite subsets the same color.

By restricting ourselves to countable sets, we will focus on the Ramsey-theoretic content of the material and not on the set-theoretic aspects, which, as we see above, can lead to "paradoxical" results.

There are a few instances in this book where we do appeal to the Axiom of Choice in the equivalent form of Zorn's Lemma, which we now state.

Zorn's Lemma. *If every chain in a partially ordered set S has an upper bound (respectively, lower bound) in S, then S contains a maximal (respectively, minimal) element.*

1.6 Exercises

1.1 Show that

$$\left(1+\frac{x}{n}\right)^n \approx e^x$$

for large n.

1.2 Let $k, n \in \mathbb{Z}^+$ be large, with $n \gg k$. Show that

$$\binom{n}{k} \approx \frac{1}{\sqrt{2\pi k}} \cdot \left(\frac{ne}{k}\right)^k.$$

1.3 Consider the k-element sets of $[1,n]$. If we color each integer in $[1,n]$ randomly with one of r colors, what is the probability that a particular k-element set is monochromatic? What is the probability that at least one of the k-element sets is monochromatic if $n > r(k-1)$?

1.4 Show that for any $n+1$ integers chosen from $\{1, 2, \ldots, 2n\}$, two of the chosen integers have the property that one divides the other.

1.5 Let $n \in \mathbb{Z}^+$ and consider a set S of n integers, none of which are divisible by n. Show that there exists a subset $\emptyset \neq T \subseteq S$ such that n divides $\sum_{t \in T} t$.

1.6 Let $n \in \mathbb{Z}^+$. Show that there are two powers of two that differ by a multiple of n.

1.7 This is a gem due to Erdős and Szekeres [70]. Let $n, m \in \mathbb{Z}^+$. Prove that every sequence of $nm + 1$ distinct numbers contains either an increasing subsequence of length $n + 1$ or a decreasing subsequence of length $m + 1$.

Hint:

For each number in the sequence let ℓ_i be the length of the longest increasing subsequence starting at the ith term.

1.8 How many arithmetic progressions of length 3 are contained in $[1, n]$? How many of length k?

1.9 Let S be a set with $|S| = n$. Let F be the set of bijections from S to S such that for $f \in F$ we have $f(s) \neq s$ for all $s \in S$. Use the Principle of Inclusion-Exclusion to show that $|F|$ is the integer nearest $\frac{n!}{e}$.

1.10 For $t \in \mathbb{Z}_n$, let

$$\chi(t) = e^{\frac{2\pi i t}{n}},$$

where $i = \sqrt{-1}$. Prove directly that, for $t \neq 0$ we have

$$\sum_{k=0}^{n-1} \chi(kt) = 0.$$

1.11 Consider the function f defined on $\mathbb{Z}_5$ by $f(x) = x^2$. Find $\widehat{f}(t)$, the discrete Fourier transform of f, and verify (i) and (iii) of Theorem 1.9.

1.12 Consider all infinite binary strings. For two strings $s_1 s_2 s_3 \ldots$ and $t_1 t_2 t_3 \ldots$, let n be the minimal positive integer for which $s_n \neq t_n$. Show that $d(s, t) = 2^{-n}$ defines a metric on the space of infinite binary strings.

1.13 Prove that Corollary 1.7 follows from Lemma 1.6.

1.14 Let $n \in \mathbb{Z}^+$. Let S be a strict subspace of $\mathbb{Q}^n$. Define $S^\perp$ to be the orthogonal complement of S. Describe $S^\perp$ and prove that every element of $\mathbb{Q}^n$ can be written as $s + s^\perp$ for some $s \in S$ and $s^\perp \in S^\perp$.

1.15 Let p be prime and let $n \in \mathbb{Z}^+$. How many $x \in \{1, 2, \ldots, p-1\}$ satisfy $x^n \equiv 1 \pmod{p}$?

1.16 Let p be prime and let $\mathbb{Z}_p^* = \mathbb{Z}_p \setminus \{0\}$ be the group under multiplication. Show that $\{i^n : i \in \mathbb{Z}_p^*\}$ is a subgroup of $\mathbb{Z}_p^*$ and determine its size. With $p = 11$ and $n = 5$, find all of its cosets. Also find the cosets when $p = 13$ and $n = 4$.

1.17 Let group G act on set S. Let $s \in S$. Prove that the stabilizer of s (see Definition 1.11) is a subgroup of G.

1.18 We say that a topological space X is endowed with the discrete topology if every point of X is an open set. Let X and Y be topological spaces each endowed with the discrete topology. If $f : X \to Y$ is a bijection, prove that f is a homeomorphism.

1.19 Let X be a compact topological space. Let S be a closed subset of X. Prove that S is compact.

1.20 Let X be a compact topological metric space. Let $B_1 \supseteq B_2 \supseteq B_3 \supseteq \cdots$ be an infinite chain of non-empty closed subsets. Prove that $\bigcap_{i=1}^{\infty} B_i$ is non-empty. This result is called Cantor's Intersection Theorem.

2

Integer Ramsey Theory

That's when 5 would get you 10 before it took you 8 just to get you 1.

–Bob Mould

Because the integers are such a simple structure, restricting Ramsey theory to the set of integers makes many results easier to digest. As such, this is a good starting point.

The author, together with Bruce Landman [129], has written a book devoted to this subject at a level suitable for undergraduates.† In this chapter, we will cover some of the same topics, but will also present more advanced ones that do not appear in [129].

2.1 Van der Waerden's Theorem

The most well-known result in Ramsey theory on the integers is van der Waerden's Theorem [205] concerning arithmetic progressions.

Theorem 2.1 (van der Waerden's Theorem). *Let $k, r \in \mathbb{Z}^+$. There exists a minimal positive integer $w(k;r)$ such that for any $n \geq w(k;r)$ the following holds: for any r-coloring of $[1,n]$ there exist $a, d \in \mathbb{Z}^+$ such that*

$$a, a+d, a+2d, \ldots, a+(k-1)d$$

is a monochromatic k-term arithmetic progression.

This theorem, along with Ramsey's Theorem (Theorem 3.6), are the most famous of Ramsey-type theorems. Van der Waerden [205] had credited Baudet with the conjecture that Theorem 2.1 is true (the translated title of van der

†At this same level, but for Ramsey theory more generally, see the beautifully illustrated short introduction given by Jungić [116].

Waerden's article is "Proof of a Baudet Conjecture"). However, extensive historical research done by Soifer is presented in his fascinating book [191] and provides compelling evidence that Baudet and Schur (who we will meet in Section 2.2.1) independently made the same conjecture.

The proof we present here is elementary. Although these proofs are easy to find (see, e.g., [83, 84, 119, 129]) this book would be incomplete without presenting one such proof. Other "non-elementary" proofs have been done as we will see in subsequent chapters of this book.

Proof of Theorem 2.1. We will prove that $w(k;r)$ exists by inducting on k, with $w(2;r) = r+1$ being the base case, which holds by the pigeonhole principle. Hence, we may assume that $w(k-1;r)$ exists for all r to prove that $w(k;r)$ exists for all r.

We will say that a set of monochromatic $(k-1)$-term arithmetic progressions $a_i - \ell d_i$, $0 \leq \ell \leq k-2$, are *end-focused* if they are each of a different color and $a_i + (k-1)d_i = a_j + (k-1)d_j$ for all i and j; that is, when each of the arithmetic progressions is extended one more term, those last terms all coincide.

We will prove the induction step by showing that for any $s \in \mathbb{Z}^+$ with $s \leq r$ there exists an integer $n = n(k,s;r)$ such that every r-coloring of $[1,n]$ either contains a monochromatic k-term arithmetic progression or contains s end-focused $(k-1)$-term arithmetic progressions. To see that the existence of $n(k,s;r)$ for all $s \leq r$ proves the existence of $w(k;r)$, note that for $n(k,r;r)$, one of the end-focused arithmetic progressions of length $k-1$ extends to a monochromatic k-term arithmetic progression (since all end-focused arithmetic progressions have different colors).

To prove the existence of $n(k,s;r)$ we induct on s, noting that $n(k,1;r) = w(k-1;r)$ works, which exists by our induction on k assumption. We now assume that $n = n(k,s-1;r)$ exists, in addition to the existence of $w(k-1,r)$ for all r. Hence, we may consider $m = 4nw(k-1;r^{2n})$.

We will show that $n(k,s;r) = m$ works. Let χ be an arbitrary r-coloring of $[1,m]$. From χ restricted to $\left[1, \frac{m}{2}\right]$ derive the r^{2n}-coloring β of $[1, w(k-1;r^{2n})]$ defined by

$$\beta(i) = \Big(\chi(2(i-1)n+1), \chi(2(i-1)n+2), \ldots, \chi(2in)\Big).$$

From the existence of $w(k-1;r^{2n})$ we have $a, a+d, \ldots, a+(k-2)d$ monochromatic under β.

From the existence of $n = n(k,s-1;r)$ and the fact that arithmetic progressions are unaffected by translation, under χ the interval $[2(a-1)n+1, (2a-1)n]$ either contains a monochromatic k-term arithmetic progression or $(s-1)$ end-focused $(k-1)$-term arithmetic progressions. We may assume the latter holds and let

$$b_j, b_j + d_j, \ldots, b_j + (k-2)d_j, \quad 1 \leq j \leq s-1,$$

be the end-focused progressions, all focused on f. Note that $f \in [2(a-1)n+$

$1, 2an]$. Next, by the definition of β, the intervals $[2(a+jd)n-2n+1, 2(a+jd)n]$, $0 \leq j \leq s-1$, are identically colored so that

$$f, f + 2dn, f + 4dn, \ldots, f + 2(k-2)dn$$

is a monochromatic (under χ) arithmetic progression with a color other than the end-focused progressions (since we are assuming no end-focused arithmetic progression extends to a monochromatic k-term arithmetic progression). Lastly, again by the definition of β, we see that for each $j \in \{1, 2, \ldots, s-1\}$, the $(k-1)$-term arithmetic progression

$$b_j, b_j + d_j + 2dn, b_j + 2d_j + 4dn, \ldots, b_j + (k-2)(d_j + 2dn)$$

is monochromatic under χ. Furthermore, each of these $s-1$ progressions is a unique color and is end-focused on $f + 2(k-1)dn$. By choice of m we have $f + 2(k-1)dn \leq m$. Hence, we have s end-focused $(k-1)$-term arithmetic progressions in any r-coloring of $[1, m]$, thereby finishing the proof. □

A nice consequence of Theorem 2.1 is that we can also require any positive multiple of the common difference d to have the same color as the arithmetic progression. We will use the Compactness Principle implicitly in the proof.

Corollary 2.2. *Let $c, k, r \in \mathbb{Z}^+$. There exists a minimal positive integer $\widehat{w}(k, c; r)$ such that for any $n \geq \widehat{w}(k, c; r)$ the following holds: for any r-coloring of $[1, n]$ there exist $a, d \in \mathbb{Z}^+$ such that $a, a+d, a+2d, \ldots, a+(k-1)d$ and cd all have the same color.*

Proof. We induct on the number of colors r. Clearly the result holds for $r = 1$, so we assume that $\widehat{w}(k, c; r-1)$ exists. Let χ be an arbitrary r-coloring of $\mathbb{Z}^+$. By Theorem 2.1, χ admits a monochromatic, say, red, arithmetic progression of length $(k-1)\widehat{w}(k, c; r-1) + 1$. Let

$$a, a + d, a + 2d, \ldots, a + (k-1)\widehat{w}(k, c; r)d$$

be such a progression. Then for each $j \in [1, \widehat{w}(k, c; r-1)]$ we see that the k-term arithmetic progression $a, a+jd, a+2jd, \ldots, a+(k-1)jd$ is red. Hence, if cjd is red for any $j \in [1, \widehat{w}(k, c; r-1)]$ we are done. Otherwise, $cd[1, \widehat{w}(k, c; r-1)]$ is $(r-1)$-colored. By the inductive hypothesis there exists $e, f \in \mathbb{Z}^+$ such that

$$e, e + f, \ldots, e + (k-1)f, cf \in [1, \widehat{w}(k, c; r-1)]$$

all have the same color when dilated by cd. Hence,

$$cde, cde + cdf, \ldots, cde + (k-1)cdf, c(cdf)$$

is the monochromatic structure we desire. □

Definition 2.3. The numbers $w(k; r)$ in Theorem 2.1 are referred to as *van der Waerden numbers.*

Van der Waerden numbers, and Ramsey-type numbers generally, are quite difficult to calculate except for the very small cases. It is easy to see that $w(2;r) = r+1$ and showing $w(3;2) = 9$ by hand is not hard. With quite a bit more work we can get $w(3;3) = 27$ and $w(4;2) = 35$ by hand. However, showing that $w(3;4) = 76$, $w(5;2) = 178$, $w(4;3) = 293$, and $w(6;2) = 1132$ requires a computer or, as in the case of the last two numbers, many computers [124, 125].

As of this writing, the above values are the only known values of $w(k;r)$. Hence, we turn to bounds. The current knowledge is not that great here, either (this is a statement about the difficulty of determining these numbers and not about the work being done on them). The best bounds are given in the following theorem.

Theorem 2.4. *For some absolute constant $c > 0$,*

$$cr^{k-1} \leq w(k;r) \leq 2^{2^{r^{2^{2^{k+9}}}}}.$$

The lower bound is due to Kozik and Shabanov [126]; the upper bound, while perhaps not looking as such, was a breakthrough by Gowers [82].

We can improve the lower bound in certain circumstances; however, the upper and lower bounds are still very far apart. Below, Theorem 2.5 is from [23] and is a generalization of a result of Berlekamp [20], who proved the $r = 2$ case.

Theorem 2.5. *Let $k, r \in \mathbb{Z}^+$ with $k-1$ prime. For $r \leq k-1$, we have $w(k;r) \geq (k-1)^{r-1}2^{k-1}$.*

Because of the difficulty of determining the values of $w(k;r)$, work has also been done to find the values of related functions. For example, for the numbers in Corollary 2.2, a simple modification to any standard program used to determine van der Waerden numbers will give $\widehat{w}(2,1;2) = 5, \widehat{w}(2,1;3) = 14, \widehat{w}(3,1;2) = 17$, and $\widehat{w}(4,1;2) = 161$. Of course $\widehat{w}(k,c;r)$ is more difficult to calculate than $w(k;r)$, so it is desirable for the related functions to have more easily calculable values. Toward this end, we consider the off-diagonal van der Waerden numbers:

Definition 2.6. Let $k_1, k_2, \ldots, k_r$ be positive integers. The least positive integer $n = w(k_1, k_2, \ldots, k_r)$ such that every r-coloring $[1,n] \to \{1,2,\ldots,r\}$ admits a k_i-term arithmetic progression with all elements of color i for some $i \in \{1,2,\ldots,r\}$ is called an *off-diagonal van der Waerden number.*

Existence is clear since $w(k_1, k_2, \ldots, k_r) \leq w(\max_i k_i; r)$. For values of small instances of these off-diagonal numbers see [129, 141]. In Table 2.1, we give some known 2-color van der Waerden numbers (both classical and off-diagonal).

Obtaining new numbers is now at a stage where straightforward computation is not feasible. Hence, a next step is to consider asymptotic behavior. As

TABLE 2.1
Known 2-color van der Waerden numbers $w(k,\ell)$ for $k,\ell \leq 15$

	$k \backslash \ell$	3	4	5	6	7	8	9	10	11	12	13	14	15
k	3	9	18	22	32	46	58	77	97	114	135	160	186	218
	4		35	55	73	109	146	309						
	5			178	206	260								
	6				1132									

mentioned before, the growth rate of $w(k;r)$ is challenging (to say the least). So, perhaps the off-diagonal van der Waerden numbers are more penetrable.

The obvious starting place for investigating the growth rate of off-diagonal van der Waerden numbers is $w(k,3)$. The current state of knowledge for this function is given in the next theorem.

Theorem 2.7. *For some absolute constants $c, d > 0$, we have*

$$\frac{8}{729} \cdot \left(\frac{k}{\log k}\right)^2 < w(k,3) \leq e^{ck^{1-d}}.$$

The lower bound is from [134] and uses the Lovász Local Lemma. We prove this lower bound in Chapter 6 (as an instance of Theorem 6.10). The upper bound is recent (2020) and comes from a preprint of Schoen [184].

Moving away from van der Waerden numbers, now that we have a guarantee of a single monochromatic arithmetic progression in r-colorings of $[1,n]$, a next step would be to determine the minimal number of monochromatic arithmetic progressions that must exist in any such coloring. In general, this is a difficult problem. In fact, the answer for 2 colors and 3 terms is still not settled. As of this writing, the theorem below gives the best-known bounds and is found in [157].

Theorem 2.8. *Let $V(n)$ be the minimal number of monochromatic 3-term arithmetic progressions over all 2-colorings of $[1,n]$. We have*

$$\frac{229475}{4489216} n^2(1+o(1)) \leq V(n) \leq \frac{239616}{4489216} n^2(1+o(1)).$$

Roughly,

$$\liminf_{n\to\infty} \frac{V(n)}{n^2} \in (.0511169, .0533760).$$

The proof of Theorem 2.8 relies on computer programs once some preliminary work is done. As the preliminary work involves a useful technique, we will present that next.

For a given 2-coloring $R \sqcup B = [1, n]$, where R is the set of red integers and B is the set of blue integers, let $V(n; R, B)$ be the set of monochromatic 3-term arithmetic progressions present.

For the remainder of this subsection, let $i = \sqrt{-1}$ and denote the complex conjugate of $c \in \mathbb{C}$ by $\overline{c}$. Noting that 3-term arithmetic progressions are solutions to $x + y - 2z = 0$, we start with (see Section 1.3.2)

$$\int_0^1 e^{2\pi i n x}\, dx = \begin{cases} 1 & \text{if } n = 0; \\ 0 & \text{otherwise,} \end{cases}$$

and define

$$f_r(x) = \sum_{k \in R} e^{2\pi i k x} \quad \text{and} \quad f_b(x) = \sum_{k \in B} e^{2\pi i k x}$$

so that

$$2V(n, R, B) = \int_0^1 \left(f_r^2(x)\overline{f_r(2x)} + f_b^2(x)\overline{f_b(2x)} \right) dx + O(n).$$

(The integral counts (x, y, z) and (y, x, z) as different solutions to $x + y = 2z$, hence we are double counting 3-term arithmetic progressions by using the integral. The $O(n)$ term is to account for solutions with $x = y = z$, which are undesired.) We rewrite the integrand as

$$(f_r(x) + f_b(x))^2(\overline{f_r(2x)} + \overline{f_b(2x)}) - (f_r(x)\overline{f_b(2x)} + f_b(x)\overline{f_r(2x)})(f_r(x) + f_b(x)) \\ - f_r(x) f_b(x)(\overline{f_r(2x)} + \overline{f_b(2x)}).$$

Doing this algebraic manipulation allows us to consider the following sets to calculate $V(n; R, B)$:

$$\begin{aligned} 2V(n; R, B) &= |\{(x, y, z) \in [1, n]^3 : x + y - 2z = 0\}| \\ &\quad - |\{(x, z) \in (R \times B) \cup (B \times R) : 2z - x \in [1, n]\}| \\ &\quad - |\{(x, y) \in R \times B : x + y \text{ is even}\}|. \end{aligned}$$

From here we now have more manageable sets to enumerate; for details the reader is referred to [157] as well as [34].

There are many other functions related to van der Waerden numbers. The interested reader may consult [129] and the references therein.

We continue this section with an old result due to Hilbert.

2.1.1 Hilbert's Cube Lemma

Arguably the first Ramsey-type result is due to Hilbert [108] as it adheres to the partitioning ethos of Ramsey theory. It should be noted that Hilbert's goal in using this lemma was a result about the irreducibility of polynomials; see [209] for more information. To describe the cube referenced in the title of this subsection, we make the following definition.

Definition 2.9 (Finite sums)**.** The set of *finite sums* of integers $x_1, x_2, \ldots, x_n$ is denoted and defined as

$$FS(x_1, x_2, \ldots, x_n) = \left\{ \sum_{i \in I} x_i : \emptyset \neq I \subseteq [1, n] \right\}.$$

Note that in the above definition, the integers x_i are not necessarily distinct.

We may now define the aforementioned cube.

Definition 2.10 (*d-cube*)**.** The *d-cube* of integers $c, x_1, x_2, \ldots, x_d$ is the set of integers $c + FS(x_1, x_2, \ldots, x_n)$.

Before getting to Hilbert's result, a little bit of intuition into why this is named the cube lemma is warranted. Consider the unit cube in $\mathbb{R}^3$ with vertices $\{(\epsilon_1, \epsilon_2, \epsilon_3) : \epsilon_i \in \{0, 1\}\}$. Let $c = (0, 0, 0), x_1 = (1, 0, 0), x_2 = (0, 1, 0)$, and $x_3 = (0, 0, 1)$. Then the other vertices are $x_1 + x_2, x_1 + x_3$, and $x_1 + x_2 + x_3$. In other words, the vertices of the unit cube are $c + FS(x_1, x_2, x_3)$. This is easily abstracted to higher dimensions and different side lengths.

We now state Hilbert's result.

Lemma 2.11 (Hilbert's Cube Lemma)**.** *Let $r, d \in \mathbb{Z}^+$. Any r-coloring of $\mathbb{Z}^+$ admits a monochromatic d-cube.*

This lemma follows from van der Waerden's Theorem since a $(d+1)$-term arithmetic progression $a, a + \ell, \ldots, a + d\ell$ is a d-cube of $a, \ell, \ell, \ldots, \ell$; however, we will give an independent proof.

Proof. We induct on d, with $d = 1$ being trivial since we only need 2 integers of the same color. Assume the result for $d - 1$. By the Compactness Principle there exists an integer $h = h(d-1; r)$ such that any r-coloring of $[1, h]$ admits a monochromatic $(d-1)$-cube. Consider any r-coloring of $[1, (r^h + 1)h]$. There are r^h possible colorings of any interval of length h and we have $r^h + 1$ disjoint intervals of length h, namely, $[(i-1)h + 1, ih]$ for $1 \leq i \leq r^h + 1$. Hence, two such intervals, say $[(a-1)h + 1, ah]$ and $[(b-1)h + 1, bh]$, with $a < b$, are colored identically.

We claim that any r-coloring of a translated interval of length h also admits a monochromatic $(d-1)$-cube. Via the obvious bijection between $[1, h]$ and $[n + 1, n + h]$, the color of any $c + \sum_{i \in I} x_i$ in $[1, h]$ is the same as the color of $(c + n) + \sum_{i \in I} x_i$ in $[n + 1, n + h]$. Hence, if $c + FS(x_1, x_2, \ldots, x_{d-1})$ is a monochromatic $(d-1)$-cube in $[1, h]$ then $(c + n) + FS(x_1, x_2, \ldots, x_{d-1})$ is a monochromatic $(d-1)$-cube in $[n + 1, n + h]$.

With this translation invariance, we apply the inductive assumption to see that the coloring of $[(a-1)h + 1, ah]$ admits a monochromatic $(d-1)$-cube of $c, x_1, x_2, \ldots, x_{d-1}$. Now consider the d-cube of $c, x_1, x_2, \ldots, x_{d-1}, (b-a)h$. Because $[(a-1)h + 1, ah]$ and $[(b-1)h + 1, bh]$ are identically colored, we see that $c + FS(x_1, x_2, \ldots, x_{d-1}, (b-a)h)$ is a monochromatic d-cube. This completes the induction. □

Remark. The above proof can easily be modified to demand that the integers $c, x_1, \ldots, x_d$ of a monochromatic d-cube be distinct by requiring $b > 2a$. This can be accomplished by considering colorings of $[1, (3r^h + 1)h]$ so that we have at least 4 identically colored intervals. In order for the d-cube to not have distinct elements, we must have $b \in \{2a-1, 2a\}$ in the proof of Lemma 2.11. By having 4 identically colored intervals, we can choose $b \notin \{2a-1, 2a\}$. Of course, this also follows from van der Waerden's Theorem by noting that the d-cube of $a, \ell, 2\ell, 3\ell, \ldots, d\ell$ is the arithmetic progression $a, a+\ell, \ldots, a+\binom{d+1}{2}\ell$.

It is remarkable that not only do we see van der Waerden's Theorem related to Hilbert's result, but material on finite sums, which we will visit later in this chapter, is also related to Hilbert's result. As we have come to understand, Hilbert had great foresight.

2.1.2 Deuber's Theorem

A useful result with somewhat the same flavor as Hilbert's Cube Lemma was given by Deuber [56] in 1973 via repeated application of van der Waerden's Theorem. In order to state the result, we require a definition.

Definition 2.12 ((m, p, c)-set)**.** Let $m, p, c \in \mathbb{Z}^+$. A set $M \subseteq \mathbb{Z}^+$ is called an *(m, p, c)-set* if there exist generators $x_0, x_1, x_2, \ldots, x_m \in \mathbb{Z}^+$ such that

$$M = \bigcup_{i=0}^{m} \left\{ cx_i + \sum_{j=i+1}^{m} \lambda_j x_j : \lambda_j \in \mathbb{Z} \cap [-p, p] \text{ for } 1 \leq j \leq m \right\},$$

where we take the empty sum to equal 0.

Note that we have the integers λ_j taking on negative values but that the set M must be a set of positive integers.

We may now state Deuber's result. Notice how this result shows the unbreakable property of (m, p, c)-sets.

Theorem 2.13 (Deuber's Theorem)**.** *Let $m, p, c, r \in \mathbb{Z}^+$ be fixed. Then there exist $M, P, C \in \mathbb{Z}^+$ so that every r-coloring of an arbitrary (M, P, C)-set admits a monochromatic (m, p, c)-set.*

We will not need the full strength of Deuber's Theorem; rather, the following weaker version (with a more digestible proof) will be useful for us.

Theorem 2.14 (Weak Deuber's Theorem)**.** *Let $m, p, c, r \in \mathbb{Z}^+$ be fixed. Every r-coloring of $\mathbb{Z}^+$ admits a monochromatic (m, p, c)-set.*

Proof. We induct on m with c, p, and r arbitrary. Consider $m = 1$. Apply Corollary 2.2 with the same c and restricted to the coloring of $c\mathbb{Z}^+$. Then we find an arithmetic progression $ca, c(a+d), c(a+2d), \dots, c(a+2pd)$ together with $c(cd)$ all of the same color. Let

$$x_0 = a + pd \qquad \text{and} \qquad x_1 = cd$$

be generators of the $(1,p,c)$-set M. Then $M = \{ca, c(a+d), \dots, c(a+2pd), c(cd)\}$ is monochromatic, completing the base case of the induction. We now assume the result for (m,p,c)-sets and will show that the result holds for $(m+1,p,c)$-sets. We start by showing that every r-coloring of $\mathbb{Z}^+$ admits a monochromatic (m,p,c)-set M such that

$$M' = \{\lambda d + m : \lambda \in \mathbb{Z} \cap [-p,p],\ m \in M\} \tag{2.1}$$

is monochromatic with $M \subseteq M' \subseteq \mathbb{Z}^+$ for some $d > 0$. We call sets like this $(m+1,p,c)$-*good* since the inclusion of the element cd would create an $(m+1,p,c)$-set.

`Claim 1.` Let $m, p, c \in \mathbb{Z}^+$ and assume that every r-coloring of $\mathbb{Z}^+$ admits a monochromatic (m,p,c)-set. Then every r-coloring of $\mathbb{Z}^+$ admits a monochromatic $(m+1,p,c)$-good set.

By the claim's hypothesis along with the Compactness Principle, there exists $n = n(m,p,c;r)$ such that any r-coloring of $[1,n]$ admits a monochromatic (m,p,c)-set.

Let χ be an arbitrary r-coloring of $\mathbb{Z}^+$. We derive an r^n-coloring of $\mathbb{Z}^+$ by defining $\widehat{\chi}(i) = (\chi(i), \chi(2i), \dots, \chi(ni))$. By van der Waerden's Theorem, $\widehat{\chi}$ admits a monochromatic $(2pn^{(2p+1)^{m+1}} + 1)$-term arithmetic progression. For convenience, denote this arithmetic progression by

$$a, a \pm d, a \pm 2d \dots, a \pm pn^{(2p+1)^{m+1}} d.$$

The color, under $\widehat{\chi}$, of this progression is $(\chi(a), \chi(2a), \dots, \chi(na))$, i.e., the r-coloring of $a[1,n]$ under χ. By the definition of n, we have a monochromatic (m,p,c)-set in $[1,n]$, under χ. Since the family of (m,p,c)-sets is closed under dilation, we have, under χ, a monochromatic (m,p,c)-set M in $a[1,n]$.

Returning to the coloring $\widehat{\chi}$ of our arithmetic progression we see that for each $s_i \in M$ we have $s_i = ak_i$ for some $k_i \in [1,n]$. This means that all ak_i have the same color under χ, so that by the definition of $\widehat{\chi}$ we have, for each $j \in [1, pn^{(2p+1)^{m+1}}]$, that $\chi((a \pm jd)k_i) = \chi(ak_i) = \chi(s_i)$.

Next, note that any (m,p,c)-set has

$$\sum_{i=0}^{m} (2p+1)^i < (2p+1)^{m+1}$$

elements. Hence, we must have less that $(2p+1)^{m+1}$ different k_i's. Setting $k = \prod_i k_i$ we see that $k < n^{(2p+1)^{m+1}}$.

Lastly, since we are considering $s_i \pm jdk_i$ for $1 \le j \le n^{(2p+1)^{m+1}}$, we have

$$\{s_i + jdk : j \in [-p,p]\} \subseteq \left\{s_i + jdk_i : j \in [-pn^{(2p+1)^{m+1}}, pn^{(2p+1)^{m+1}}]\right\}.$$

As this holds for each $s_i \in M$ we have $\chi(s_i) = \chi(s_i \pm j(dk))$ for all $j \in [1,p]$. Hence,

$$M' = \{m + j(dk) : m \in M, j \in [-p,p]\}$$

is a monochromatic $(m+1,p,c)$-good set, which finishes the proof of Claim 1. ◇

To finish the proof, we will use the existence of monochromatic $(m+1,q,c)$-good sets for all q to extend to monochromatic $(m+1,p,c)$-sets.

`Claim 2.` Fix $m,p,c \in \mathbb{Z}^+$. Assume every r-coloring of $\mathbb{Z}^+$ admits a monochromatic $(m+1,q,c)$-good set for any $q \in \mathbb{Z}^+$. Then every r-coloring of $\mathbb{Z}^+$ admits a monochromatic $(m+1,p,c)$-set.

To prove the claim we induct on r, with $r = 1$ being trivial. We assume the result holds for $r-1$ colors. Applying the Compactness Principle to the inductive assumption, there exists an integer $n = n(m+1,p,c;r-1)$ such that every $(r-1)$-coloring of $[1,n]$ admits a monochromatic $(m+1,p,c)$-set.

Consider an arbitrary r-coloring of $\mathbb{Z}^+$. By the claim's assumption, there exists a monochromatic, say red, $(m+1,np,c)$-good set M'. Using the notation from Equation (2.1), if any jcd, with $j \in [1,n]$, is red, then $M' \cup \{jcd\}$ contains a red $(m+1,p,c)$-set and we are done. Otherwise $cd[1,n]$ is $(r-1)$-colored. By the definition of n, we have a monochromatic $(m+1,p,c)$-set M in $[1,n]$. Noting that cdM is also an $(m+1,p,c)$-set, we are done with Claim 2. ◇

The theorem now follows from Claims 1 and 2 via induction on m. □

2.2 Equations

We move on from arithmetic progressions and (m,p,c)-sets to solutions of equations. We start by noting that the 3-term arithmetic progression $a, a+d, a+2d$ provides a solution to $x+y=2z$. Hence, by the existence of $w(3;r)$ we see that every r-coloring of $\mathbb{Z}^+$ admits a monochromatic solution to $x+y=2z$ (with $x \neq y$). Are there other equations that also have this Ramsey property? In this section we explore this question.

2.2.1 Schur's Theorem

The simplest of nontrivial equations is probably $x+y=z$ (clearly, $x=y$ has the Ramsey property, while it is easy to show that $y = x+b$, $b \neq 0$, and $y = cx$, $c \neq 0,1$, do not). By Corollary 2.2, the existence of $\widehat{w}(2,1;r)$ shows that we do have a monochromatic solution to $x+y=z$ with $x=a$ and $y=d$. However, such a monochromatic solution was proven by Schur [186] before van der Waerden's Theorem was proven. We follow Schur's original proof.

Theorem 2.15 (Schur's Theorem). *Let $r \in \mathbb{Z}^+$. There exists a minimal positive integer $s = s(r)$ such that every r-coloring of $[1, s]$ contains a monochromatic solution to $x+y=z$.*

Proof. We will show that $s(r) < er!$, thereby proving existence. Let $n_0 \in \mathbb{Z}^+$ be arbitrary and assume that we have an r-coloring of $[1, n_0]$ with no monochromatic solution to $x+y=z$. Some color occurs at least $\frac{n_0}{r}$ times. Let this be color 1 and let $x_1, x_2, \ldots, x_{n_1}$ be all of the integers of color 1, with $n_1 \geq \frac{n_0}{r}$. By assumption,

$$S_1 = \{x_k - x_1 : 2 \leq k \leq n_1\}$$

is void of color 1.

Within S_1, some color occurs at least $\frac{n_1-1}{r-1}$ times. Let this be color 2 (it cannot be color 1). Let $y_1, y_2, \ldots, y_{n_2}$, with $n_2 \geq \frac{n_1-1}{r-1}$, be all of the integers of color 2. By assumption,

$$S_2 = \{y_k - y_1 : 2 \leq k \leq n_2\}$$

is void of color 2. Since $y_j = x_{k_j} - x_1$ we see that $y_i - y_1 = x_{k_i} - x_{k_1}$ so that $x_{k_1} + (y_i - y_1) = x_{k_i}$. Hence, S_2 must also be void of color 1.

We continue in this manner to define sets S_i of size $n_i - 1$, where $n_i \geq \frac{n_{i-1}-1}{r-i+1}$, that are void of colors $1, 2, \ldots, i$. Hence, we have $n_0 \leq rn_1$ and $n_i \leq (r-i+1)n_{i+1} + 1$ for $1 \leq i \leq k$ for some $k < r$ (since S_r must be void of all colors).

Stringing these together, we have

$$\begin{aligned} n_0 \leq rn_1 \leq r((r-1)n_2 + 1) &= r + r(r-1)n_2 \\ &\leq r + r(r-1) + r(r-1)(r-2)n_3 \\ &\vdots \\ &\leq \sum_{i=0}^{r} \frac{r!}{i!} \\ &< r! \sum_{i=0}^{\infty} \frac{1}{i!} = er!, \end{aligned}$$

thereby proving the theorem. □

Definition 2.16 (Schur number, Schur triple). The numbers $s(r)$ in Theorem 2.15 are called *Schur numbers*, while solutions to $x+y=z$ are called *Schur triples*.

TABLE 2.2
Known Schur numbers $s(r)$

r	1	2	3	4	5
$s(r)$	1	5	14	45	161

As simply as these numbers are defined, we know the value of only 5 Schur numbers, given in Table 2.2.

The value of $s(4)$ was determined by Baumert [10] in 1961 with the aid of a computer. The last value was only recently (2018) determined by Heule [105] using a massively parallel program and an enormous (2 pentabytes) amount of storage.

The original upper bound of $er!$ has not been significantly improved. The current best upper bound of $\left(e - \frac{1}{6}\right) r!$ can be found in [217]. For a lower bound, we again turn to Schur.

Theorem 2.17. *Let* $r \in \mathbb{Z}^+$. *Then* $s(r+1) \geq 3s(r) - 1$.

Proof. Let $n = s(r)$ and let $\chi : [1, n-1] \to \{1, 2, \ldots, r\}$ be an r-coloring that avoids monochromatic solutions to $x + y = z$. We will extend χ to an $(r+1)$-coloring of $[1, 3n-2]$. Let $\widehat{\chi}$ be this extended coloring defined as

$$\begin{aligned} \widehat{\chi}(i) &= \chi(i) && \text{for } i \in [1, n-1]; \\ \widehat{\chi}(i) &= r+1 && \text{for } i \in [n, 2n-1]; \\ \widehat{\chi}(i) &= \chi(i - (2n-1)) && \text{for } i \in [2n, 3n-2]. \end{aligned}$$

Clearly we have no monochromatic solution to $x + y = z$ of color $r + 1$ or with $z \leq n - 1$. Hence, the only possibility is with $z \in [2n, 3n-2]$. We may assume $x \leq y$ in this situation so that $y \in [2n, 3n-2]$, too. However, if $x + y = z$ is a monochromatic solution then, by the way $\widehat{\chi}$ is defined, so too is $x + (y - (2n-1)) = z - (2n-1)$, a contradiction. □

We may use Theorem 2.17 to give a lower bound of

$$s(r) \geq \frac{1}{2}(3^r + 1)$$

via an easy induction on r. However, a result of Abbott and Hanson [1] coupled with the determination $s(5) = 161$ by Heule [105] allows us to deduce that $s(r+5) \geq 323s(r) + 161$ and conclude (using Fekete's Lemma; see Corollary 1.7) that

$$\lim_{r\to\infty} (s(r))^{\frac{1}{r}} \geq (323)^{\frac{1}{5}} > 3.1757.$$

As we explored in Section 2.1, it is natural to consider the minimal number of monochromatic Schur triples over 2-colorings of $[1, n]$. The situation in this case is better than for 3-term arithmetic progressions: the minimum has been determined.

Theorem 2.18. *The minimal number of monochromatic Schur triples over 2-colorings of* $[1, n]$ *is*

$$\frac{1}{22}n^2(1 + o(1)).$$

Theorem 2.18 was independently proved in [171] and [183] (see also [129]). For generalizations to triples satisfying $x + ay = z$, the minimum has been determined to be

$$\frac{1}{2a(a^2 + 2a + 3)}n^2(1 + o(1))$$

by Thanatipanonda and Wong [200].

2.2.2 Rado's Theorem

Thus far we have seen that the equations $x+y = z$ and $x+y = 2z$ each admit monochromatic solutions under any finite coloring of $\mathbb{Z}^+$. On the other hand, it is easy to see that by coloring the entire intervals $[2^i, 2^{i+1} - 1]$ with color i modulo 2, we have no guaranteed monochromatic solution to $x = 2y$ even for 2 colors. In order to discuss these types of results, we make the following definition.

Definition 2.19 (r-regular, regular)**.** Let $r \in \mathbb{Z}^+$. We say that an equation $\mathcal{E}$ is *r-regular* if every r-coloring of $\mathbb{Z}^+$ admits a monochromatic solution to $\mathcal{E}$. If $\mathcal{E}$ is r-regular for all $r \in \mathbb{Z}^+$, then we say that $\mathcal{E}$ is *regular.*

In this subsection we will limit our attention to linear homogeneous equations. We start with a lesser-known result of Rado [163], who was a student of Schur.

Theorem 2.20. *Let $n \in \mathbb{Z}^+$ with $n \geq 3$ and let c_i, $1 \leq i \leq n$, be non-zero integers. The equation $\sum_{i=1}^{n} c_i x_i = 0$ is 2-regular if for some i_1 and i_2 we have $c_{i_1} > 0$ and $c_{i_2} < 0$.*

Proof. By equating variables as necessary, it suffices to prove that every 2-coloring of $\mathbb{Z}^+$ admits a monochromatic solution to $ax + by = cz$ for any $a, b, c \in \mathbb{Z}^+$.

Assume, for a contradiction, that there exists a 2-coloring of $\mathbb{Z}^+$ with no monochromatic solution. We may assume that every dilation of $\mathbb{Z}^+$, say $d\mathbb{Z}^+$, contains elements of both colors; otherwise, (dbc, dac, dab) would be a monochromatic solution and we would be done.

Let ac be red and let acj be blue ($j \geq 2$) as both are members of the dilation $ac\mathbb{Z}^+$. Within $bc(j-1)\mathbb{Z}^+$ there exists a blue element. Let this element be t_1 and let

$$t_1 = bc(j - 1)k.$$

Consider

$$u_1 = ab(j-1)k + abj.$$

Since (t_1, acj, u_1) is a solution, we may assume that u_1 is red. Now consider

$$t_2 = bc(j-1)(k+1).$$

Since (t_2, ac, u_1) is a solution, we may assume that t_2 is blue.

We proceed in this fashion (starting with $u_2 = ab(j-1)(k+1) + abj$) to deduce that

$$t_i = bc(j-1)(k+i-1)$$

is blue for all $i \in \mathbb{Z}^+$. Since $\{t_i\}_{i\in\mathbb{Z}^+} \supseteq bc(j-1)k\mathbb{Z}^+$, we have a monochromatic dilation of $\mathbb{Z}^+$, contradicting our assumption. □

While this theorem tells us which equations are 2-regular, it does not address the more interesting question of which equations are regular. For example, consider the 3-coloring $\chi : \mathbb{Z}^+ \to \{0, 1, 2\}$ with $\chi(i)$ equal to a modulo 3, where a is the exponent of 2 in the prime decomposition of i. Then χ does not admit a monochromatic solution to $x + 2y = 4z$ (the justificiation for this is left to the reader as Exercise 2.7). Hence, this equation is 2-regular, but not 3-regular.

In fact, Alexeev and Tsimerman [3] have shown that for every r there exists an equation that is $(r-1)$-regular but not r-regular. This equation has r variables (so that the number of variables increases with the number of colors), which means that it does not answer the following conjecture of Rado.

Conjecture 2.21 (Rado's Boundedness Conjecture)**.** For each $k \geq 3$, there exists a minimal positive constant $c = c(k)$ such that if a linear homogeneous equation in k variables is c-regular, then it is regular.

The first case of this conjecture has been settled by Fox and Kleitman [75] who showed that $c(3) \leq 24$. Note that Alexeev and Tsimerman's equation shows that $c(k) \geq k$.

Although Rado's Boundedness Conjecture has not been resolved, Rado did provide us with a complete description of which equations are regular.

Theorem 2.22. *Let $n \geq 3$ and let $c_i, 1 \leq i \leq n$, be non-zero integers. The equation*

$$\sum_{i=1}^{n} c_i x_i = 0$$

is regular if and only if for some non-empty subset $S \subseteq [1, n]$ we have

$$\sum_{s \in S} c_s = 0.$$

Theorem 2.22 follows immediately from the more general result (Theorem 2.26, below) that Rado [162, 163, 164] proved concerning systems of linear equations. In order to state the regularity criteria for linear homogeneous systems proven by Rado, we need a definition.

Definition 2.23 (Columns Condition). Let A be an $n \times m$ matrix with column vectors $\mathbf{c}_1, \mathbf{c}_2, \ldots, \mathbf{c}_m$. We say that A satisfies the *columns condition* if, after renumbering if necessary, there exist indices $i_0 = 0 < i_1 < i_2 < \cdots < i_k = m$ such that, for

$$\mathbf{s}_j = \sum_{i=i_{j-1}+1}^{i_j} \mathbf{c}_i, \qquad 1 \leq j \leq k,$$

the following hold:

(i) $\mathbf{s}_1 = \mathbf{0} \in \mathbb{Z}^n$;

(ii) for $2 \leq j \leq k$ we can write $\mathbf{s}_j$ as a linear combination over $\mathbb{Q}$ of $\mathbf{c}_1, \mathbf{c}_2, \ldots, \mathbf{c}_{i_{j-1}}$.

Before getting to Rado's Theorem, we note the following linear algebra results relating solutions to $Ax = \mathbf{0}$ to the columns condition. For the definition of (m, p, c)-sets, see Definition 2.12.

Lemma 2.24. *Let A be an $n \times m$ matrix that satisfies the columns condition. Then there exist positive integers p and c such that every (m, p, c)-set contains a solution to $Ax = \mathbf{0}$.*

Proof. Using the notation from Definition 2.23, condition (i) means we have

$$\sum_{i=1}^{i_1} \mathbf{c}_i = \mathbf{0},$$

while condition (ii) means that for each $j \in \{2, 3, \ldots, k\}$ we have

$$\sum_{i=1}^{i_{j-1}} q_i^{(j)} \mathbf{c}_i + \sum_{i=i_{j-1}+1}^{i_j} \mathbf{c}_i = \mathbf{0}$$

for some $q_i^{(j)} \in \mathbb{Q}$, $1 \leq i \leq i_{j-1}$. These mean that

$$(\underbrace{1, 1, \ldots, 1}_{i_1}, 0, 0, \ldots, 0) \in \mathbb{Z}^m$$

and

$$(q_1^{(j)}, q_2^{(j)}, \ldots, q_{i_{j-1}}^{(j)}, \underbrace{1, 1, \ldots, 1}_{i_j - i_{j-1}}, 0, 0, \ldots, 0) \in \mathbb{Q}^m, \qquad 2 \leq j \leq k,$$

are rational solutions to $Ax = \mathbf{0}$.

Let c be the absolute value of the product of all denominators appearing in all $q_i^{(j)}$. Multiplying each solution above by c, we now have the following integer solutions to $Ax = \mathbf{0}$:

$$(\underbrace{c, c, \ldots, c}_{i_1}, 0, 0, \ldots, 0) \in \mathbb{Z}^m$$

and

$$(z_1^{(j)}, z_2^{(j)}, \ldots, z_{i_{j-1}}^{(j)}, \underbrace{c, c, \ldots, c}_{i_j - i_{j-1}}, 0, 0, \ldots, 0) \in \mathbb{Z}^m, \qquad 2 \leq j \leq k,$$

for some $z_i^{(j)} \in \mathbb{Z}$, $1 \leq i \leq i_{j-1}$.

Using these particular solutions, we see that $\mathbf{y} = (y_1, y_2, \ldots, y_m)$ with

$$y_i = cx_1 + \sum_{\ell=2}^{k} z_i^{(\ell)} x_\ell, \qquad 1 \leq i \leq i_1,$$

and, for $2 \leq j \leq k$,

$$y_i = cx_j + \sum_{\ell=j+1}^{k} z_i^{(\ell)} x_\ell, \qquad i_{j-1} + 1 \leq i \leq i_j,$$

is a solution for any $x_i \in \mathbb{Z}^+$, $1 \leq i \leq k$. Letting $p = \max_{i,\ell} |z_i^{(\ell)}|$ and noting that $k \leq m$, we see that every (m, p, c)-set contains a solution, namely $\mathbf{y}$. □

Lemma 2.25. *Let $m \in \mathbb{Z}^+$ and $k \in \mathbb{Z}^+ \cup \{0\}$. Assume that $\mathbf{c}_1, \mathbf{c}_2, \ldots, \mathbf{c}_k, \mathbf{d} \in \mathbb{Z}^m$ are vectors such that $\mathbf{d}$ is not a linear combination over $\mathbb{Q}$ of $\mathbf{c}_1, \mathbf{c}_2, \ldots, \mathbf{c}_k$. Then there exists $n \in \mathbb{Z}^+$ such that for all primes $p > n$ we have, for any $j \in \mathbb{Z}^+ \cup \{0\}$,*

$$p^j \mathbf{d} \not\equiv \sum_{i=1}^{k} a_i \mathbf{c}_i \pmod{p^{j+1}}$$

for all $(a_1, a_2, \ldots, a_k) \in \mathbb{Z}^k$ (where we take the empty sum to be $\mathbf{0}$).

Proof. For $k = 0$, by assumption $\mathbf{d} \neq \mathbf{0}$ and taking $n > \max_i |d_i|$ works where $\mathbf{d} = (d_1, d_2, \ldots, d_m)$, so we may assume $k \in \mathbb{Z}^+$. Let $S = \text{span}(\mathbf{c}_1, \mathbf{c}_2, \ldots, \mathbf{c}_k)$ over $\mathbb{Q}$. Then $S \neq \mathbb{Q}^m$ as $\mathbf{d} \notin S$. Hence, we can find a vector $\mathbf{x} \neq \mathbf{0}$ in the orthogonal complement $S^\perp$. Since $\mathbf{d} \notin S$, we can write $\mathbf{d} = \mathbf{d}_1 + \mathbf{d}_2$ with $\mathbf{d}_1 \in S$ and $\mathbf{0} \neq \mathbf{d}_2 \in S^\perp$. Using the dot product, we have $\mathbf{x} \odot \mathbf{d} = \mathbf{x} \odot \mathbf{d}_1 + \mathbf{x} \odot \mathbf{d}_2 = 0 + \mathbf{x} \odot \mathbf{d}_2 \neq 0$. Multiplying through by the product of denominators in $\mathbf{x}$, if necessary, we can assume that $\mathbf{0} \neq \mathbf{x} \in \mathbb{Z}^m$ so that $\mathbf{x} \odot \mathbf{c}_i = 0$ for $1 \leq i \leq k$, while $\mathbf{x} \odot \mathbf{d} \neq 0$. Let $n = |\mathbf{x} \odot \mathbf{d}|$.

Assume, for a contradiction, that there exists $(b_1, b_2, \ldots, b_k) \in \mathbb{Z}^k$ such that $p^j\mathbf{d} \equiv \sum_{i=1}^{k} b_i\mathbf{c}_i \pmod{p^{j+1}}$ for some $j \in \mathbb{Z}^+ \cup \{0\}$ and some prime $p > n$. Taking the dot product of both sides with respect to $\mathbf{x}$ yields

$$\{\pm p^j n\} \ni p^j(\mathbf{x} \odot \mathbf{d}) \equiv \sum_{i=1}^{k} b_i(\mathbf{x} \odot \mathbf{c}_i) \equiv 0 \pmod{p^{j+1}},$$

giving $p \mid n$, which contradicts $p > n$ and p prime. □

We now present Rado's Theorem, which categorizes those linear homogeneous systems that are regular.

Theorem 2.26 (Rado's Theorem). *The homogeneous system of linear equations $A\mathbf{x} = \mathbf{0}$ is regular if and only if A satisfies the columns condition.*

Proof. First, assume that A satisfies the columns condition. By Lemma 2.24, there exists a solution to $Ax = \mathbf{0}$ in every (m,p,c)-set for some specific values of $m, p,$ and c. For any $r \in \mathbb{Z}^+$, by Theorem 2.14, every r-coloring of $\mathbb{Z}^+$ admits a monochromatic (m,p,c)-set for any $m,p,c \in \mathbb{Z}^+$. Hence, we have a monochromatic solution to $Ax = \mathbf{0}$ under any r-coloring of $\mathbb{Z}^+$, for any $r \in \mathbb{Z}^+$. Thus, the system is regular.

In the other direction, assume that the system is regular. We will show that A satisfies the columns condition. Let $\mathbf{c}_1, \mathbf{c}_2, \ldots, \mathbf{c}_m$ be the column vectors of A. For every $\emptyset \neq I \subseteq \{1,2,\ldots,m\}$ we apply Lemma 2.25 with $\mathbf{d} = \sum_{i\in I} \mathbf{c}_i$ to conclude that one of the following holds:

(a) $\mathbf{d}$ is a linear combination of $\{\mathbf{c}_j : j \in [1,m] \setminus I\}$ over $\mathbb{Q}$;

(b) there exists an integer $n(I)$ such that for all primes $p > n(I)$ the incongruence of Lemma 2.25 holds.

We may assume that (b) occurs at least once; otherwise, (a) holds with $I = [1,m]$ so that $\mathbf{d} = \mathbf{0}$ and the columns condition is satisfied.

Let $N = \max_I n(I)$ and let $p > N$ be prime. We now define a $(p-1)$-coloring χ of $\mathbb{Z}^+$. Each $i \in \mathbb{Z}^+$ can be written as $i = p^j r_i$ for some $r_i \not\equiv 0 \pmod{p}$. Define $\chi(i)$ to be r_i modulo p for each i.

Since $A\mathbf{x} = \mathbf{0}$ is regular, there exists a monochromatic solution $\mathbf{x} = (x_1, x_2, \ldots, x_m)$ under χ. Let the color be s so that, for $1 \leq i \leq m$, we have $x_i \equiv sp^{j_i} \pmod{p^{j_i+1}}$ for some non-negative integer j_i. By reordering the columns of A if necessary, we have $j_1 \leq j_2 \leq \cdots \leq j_m$. To account for possible equalities we find $i_0 = 0 < i_1 < i_2 < \cdots < i_k = m$ and $j_1 = n_1 < n_2 < \cdots < n_k = j_m$ so that $n_\ell = j_t$ for all $i_{\ell-1} + 1 \leq t \leq i_\ell$ for $1 \leq \ell \leq k$.

First, we show that $\sum_{i=1}^{i_1} \mathbf{c}_i = \mathbf{0}$ to satisfy columns condition (i). To see this, consider $\sum_{i=1}^{m} x_i\mathbf{c}_i$ modulo p^{n_1+1}. We have

$$\mathbf{0} = \sum_{i=1}^{m} x_i\mathbf{c}_i = \sum_{i=1}^{i_1} x_i\mathbf{c}_i + \sum_{i=i_1+1}^{m} x_i\mathbf{c}_i \equiv sp^{n_1} \sum_{i=1}^{i_1} \mathbf{c}_i + \mathbf{0} \pmod{p^{n_1+1}}$$

since $p^{n_1+1} \mid x_i$ for all $i > i_1$. We conclude, since $(s,p) = 1$, that $\sum_{i=1}^{i_1} \mathbf{c}_i \equiv \mathbf{0} \pmod{p}$. By Lemma 2.25 with

$$\mathbf{d} = \sum_{i=1}^{i_1} \mathbf{c}_i$$

and $k = 0$ we deduce that $\mathbf{d} = \mathbf{0}$.

By considering $\sum_{i=1}^{m} x_i \mathbf{c}_i$ modulo $p^{n_\ell+1}$ we similarly have

$$\mathbf{0} = \sum_{i=1}^{i_{\ell+1}} x_i \mathbf{c}_i \equiv \sum_{i=1}^{i_\ell} x_i \mathbf{c}_i + sp^{n_\ell} \sum_{i=i_\ell+1}^{i_{\ell+1}} \mathbf{c}_i \pmod{p^{n_\ell+1}}.$$

Let $\overline{s} \in \{1, 2, \ldots, p-1\}$ be such that $s\overline{s} \equiv -1 \pmod{p}$ (which is possible since $(s,p) = 1$) and multiply through by $\overline{s}$ to get

$$p^{n_\ell} \sum_{i=i_\ell+1}^{i_{\ell+1}} \mathbf{c}_i \equiv \sum_{i=1}^{i_\ell} \overline{s} x_i \mathbf{c}_i \pmod{p^{n_\ell+1}}.$$

Applying Lemma 2.25 with

$$\mathbf{d} = \sum_{i=i_\ell+1}^{i_{\ell+1}} \mathbf{c}_i$$

and $k = i_\ell$ we find that $\mathbf{d}$ is a linear combination (over $\mathbb{Q}$) of $\mathbf{c}_1, \mathbf{c}_2, \ldots, \mathbf{c}_{i_\ell}$, as needed to satisfy columns condition (ii). □

Remark. The treatment of Rado's Theorem presented above is similar to that found in [84] and in [147].

As we noted before, Theorem 2.22 follows from Theorem 2.26. To see this, consider the $1 \times m$ matrix A of the coefficients of the equation under consideration. The criteria in Theorem 2.22 is columns condition (i), while columns condition (ii) trivially holds for any equation with integer coefficients.

Rado [163] also considered inhomogeneous systems $A\mathbf{x} = \mathbf{b} \neq \mathbf{0}$ and showed that there is a monochromatic solution in any finite coloring of $\mathbb{Z}^+$ if and only if one of two conditions holds (we use the notation $\mathbf{1} = (1, 1, \ldots, 1)$):

(i) there exists $c \in \mathbb{Z}^+$ such that $cA\mathbf{1} = \mathbf{b}$; or

(ii) A satisfies the columns condition and there exists $c \in \mathbb{Z}$ such that $cA\mathbf{1} = \mathbf{b}$.

2.2.2.1 Some Rado Numbers

Before moving on to other equations, we give some results about Rado numbers. Unlike van der Waerden and Schur numbers, Rado numbers are a class of numbers covering all systems of equations (typically linear and homogeneous). No over-arching formula is known for arbitrary systems (or even an arbitrary equation). To date, the determination of these numbers has almost solely been restricted to single equations. Thus, we will restrict our definition to single equations.

Definition 2.27 (Rado numbers)**.** Let $r \in \mathbb{Z}^+$. Let $\mathcal{E}$ be an r-regular linear homogeneous equation. The minimal positive integer $n = n(\mathcal{E}; r)$ such that every r-coloring of $[1, n]$ admits a monochromatic solution to $\mathcal{E}$ is called the *r-color Rado number for $\mathcal{E}$.*

We gather, in Table 2.3, all known completely determined 2-color Rado numbers (as of this writing). As the proofs of these numbers are mostly elementary (but, perhaps, clever) color-forcing arguments, we will not provide proofs. The reader may find proofs by following the reference(s) given by each result. In Table 2.3, parameters a, a_i, b, k, and ℓ are positive integers.

Note that when restricted to two colors, Theorem 2.20 states that we need not satisfy the criteria of Theorem 2.22 (a subset of coefficients summing to 0) in order for the 2-color Rado number to exist.

Although there are known numerical values for more than two colors for Rado numbers for some specific equations, there are no formulas as we have for two colors, except for a few specific cases.

As we can see from Table 2.3, we are homing in on the determination of the 2-color Rado number for an arbitrary homogeneous linear equation.

2.2.3 Nonlinear Equations

Thanks to Rado, we have a complete understanding of which systems of linear equations are regular. An obvious next step is the investigation of nonlinear equations.

Perhaps Pythagorean triples, i.e., solutions to $x^2 + y^2 = z^2$, pop to mind first. While this is a homogeneous equation, let's not make the situation harder than we must. We'll start with $x+y = z^2$. The following result is due to Green and Lindqvist [88]; we follow Pach [156] for the proof of 2-regularity.

Theorem 2.28. *The equation $x + y = z^2$ is 2-regular, but not 3-regular.*

Proof. Of course $(x, y, z) = (2, 2, 2)$ is always a monochromatic solution under any coloring, so we do not allow this as a valid monochromatic solution.

We start by exhibiting a 3-coloring of $\mathbb{Z}^+$ with no monochromatic solution. Define the intervals

$$I_j = \{i \in \mathbb{Z}^+ : 2^j \leq i \leq 2^{j+1} - 1\} \quad \text{for} \quad j = 0, 1, 2, \ldots.$$

Let the colors be $0, 1$, and 2. For $j = 0, 1$, and 2, color all elements of I_j by color j. For $j = 3, 4, \ldots$, in order, color all elements of I_j with a color missing from

$$I_{\lfloor \frac{j}{2} \rfloor} \cup I_{\lfloor \frac{j}{2} \rfloor + 1}.$$

TABLE 2.3
Known formulas for 2-color Rado numbers

$\mathcal{E}$	$n(\mathcal{E};2)$	Ref.
$ax+ay=2z$, $(a,2)=1$	$\frac{1}{2}a(a^2+1)$	[33, 129]
$ax+ay=3z$, $a\geq 4$, $(a,3)=1$	$\frac{a}{9}(4a^2+a+3)+\begin{cases}\frac{a}{9} & \text{if } a\equiv 1 \pmod 9\\ \frac{6}{9} & \text{if } a\equiv 2 \pmod 9\\ \frac{a+6}{9} & \text{if } a\equiv 4 \pmod 9\\ \frac{3a+3}{9} & \text{if } a\equiv 5 \pmod 9\\ \frac{4a}{9} & \text{if } a\equiv 7 \pmod 9\\ \frac{3}{9} & \text{if } a\equiv 8 \pmod 9\end{cases}$	[33, 129]
$ax+ay=bz$, $b\geq 4, (a,b)=1$	$\binom{b+1}{2}$ for $a\leq\frac{b}{4}$ $\lceil\frac{b}{2}\rceil$ for $\frac{b}{4}\leq a<\frac{b}{2}$ ab for $\frac{b}{2}<a<b$ $\lceil\frac{a^2}{b}\rceil$ for $b<a$	[33, 129]
$ax+by=bz$, $(a,b)=1$	a^2+3a+1 for $b=1$ b^2 for $a<b$ a^2+a+1 for $2\leq b<a$	[33, 129]
$ax+by=(a+b)z$, $(a,b)=1$	$4(a+b)-1$ if $(a,b)\in\{(1,4k):k\in\mathbb{Z}^+\}$ or $(a,b)=(3,4)$ $4(a+b)+1$ otherwise	[95, 129]
$\sum_{i=1}^{k} a_ix_i=x_{k+1}$	$a\left(\sum_{i=1}^{k} a_i\right)^2+\sum_{i=1}^{k} a_i-a$; where $a=\min_i a_i$	[94]
$\sum_{i=1}^{k} x_i=ax_{k+1}$, $a\geq 2$, $k>2a$ (also $k=2a$ if $3\mid a$)	$\left\lceil\frac{k}{a}\left\lceil\frac{k}{a}\right\rceil\right\rceil$	[179]
$\sum_{i=1}^{k} x_i=\sum_{i=1}^{\ell} a_iy_i$	$\left\lceil\frac{k}{\ell}\left\lceil\frac{k}{\ell}\right\rceil\right\rceil$ if all $a_i=1$ and $k\geq\ell$ $\left\lceil\frac{k}{a}\left\lceil\frac{k}{a}\right\rceil\right\rceil$ for $k\geq\sum_{i=1}^{\ell}a_i$; and for $2k\geq\sum_{i=1}^{\ell}a_i$ except for some sporadic values where $a=\sum_{i=1}^{\ell}a_i$	[180]

Assume, for a contradiction, that $a+b=c^2$ with $a \leq b$ is a monochromatic solution under this coloring. For some j we have $b \in I_j$. Since $a \leq b$ we have $2^j < a+b < 2^{j+2}$ so that $2^{\frac{j}{2}} < c < 2^{\frac{j}{2}+1}$. Hence, $c \in I_{\lfloor \frac{j}{2} \rfloor} \cup I_{\lfloor \frac{j}{2} \rfloor+1}$. By construction, if $j \geq 3$, this means that the color of c and b are different. For $j \in \{0,1,2\}$ we have $b < 8$ so that $c \in \{2,3\}$. For $c=2$, the only solutions are $(a,b,c)=(1,3,2)$ and $(2,2,2)$. The first is not monochromatic and the latter is not a valid monochromatic solution. For $c=3$, the solutions are $(2,7,3),(3,6,3)$, and $(4,5,3)$, none of which are monochromatic. Hence, the equation is not 3-regular.

We continue by showing that for $n \geq 14$, the interval $[n,(10n)^4]$ admits a monochromatic solution under any 2-coloring. To this end, let $\chi : [n,(10n)^4] \to \{-1,1\}$ be an arbitrary coloring. We use the colors -1 and 1 instead of the more standard 0 and 1 since we will be deriving inequalities about sums of colors and sums of -1s and 1s have more easily described properties.

Assume, for a contradiction, that χ does not admit a monochromatic solution. Since $(x,y,z)=(n^2,80n^2,9n)$ is a solution, we see that $[9n,80n^2]$ cannot be entirely of one color. Hence, there exists

$$k \in [9n, 80n^2-1]$$

such that, without loss of generality, $\chi(k)=1$ and $\chi(k+1)=-1$.

Consider solutions (i, x^2-i, x) for $i \in [n, x^2-n]$ with $x \in \{k,k+1\}$. Since $\chi(k)=1$ and $\chi(k+1)=-1$, to avoid monochromatic solutions we must have

$$\chi(i)+\chi(k^2-i) \leq 0 \quad \text{and} \quad \chi(i)+\chi((k+1)^2-i) \geq 0.$$

The second inequality yields

$$\chi(i+2k+1)+\chi(k^2-i) \geq 0$$

for every $i \in [n,k^2-n]$. Next, note that if $\chi(i+2k+1)=-1$ then we have $\chi(k^2-i)=1$ so that $\chi(i)=-1$. Hence, we have $\chi(i) \leq \chi(i+2k+1)$ for all $i \in [n,k^2-n]$. We deduce that for each $j \in [n,n+2k]$, the color pattern of

$$\chi(j), \chi(j+2k+1), \chi(j+2(2k+1)), \ldots$$

inside of $[n,k^2-n]$ is

$$-1,-1,\ldots,-1,1,1,\ldots,1,$$

where the number of -1s or 1s may be zero.

For every $j \in [n,n+2k]$ let

$$A_j = \{j+\ell(2k+1) : \ell \in \mathbb{Z}^+ \cup \{0\}; j+\ell(2k+1) \leq k^2-n\}$$

and notice (heeding the disjoint union usage) that

$$[n,k^2-n] = \bigsqcup_{j=n}^{n+2k} A_j.$$

Define the following statistics on each A_j:

$$f(A_j) = \begin{cases} \min(x \in A_j : \chi(x) = 1) & \text{if it exists;} \\ \infty & \text{otherwise,} \end{cases}$$

and

$$g(A_j) = \begin{cases} \max(x \in A_j : \chi(x) = -1) & \text{if it exists;} \\ 0 & \text{otherwise.} \end{cases}$$

Note that when f is finite and g is non-zero, we have

$$g(A_j) = f(A_j) - (2k+1). \tag{2.2}$$

Using these statistics, define functions on $[n, k^2 - n]$ by $f(x) = f(A_i)$ and $g(x) = g(A_i)$, where A_i is the unique arithmetic progression containing x.

As we have $f(i) \geq i$ for any $i \in [n, n+2k]$, in the situation that $f(j) = j$ and $f(k^2 - j) \leq k^2 - j$ we see that both j and $k^2 - j$ have color 1 so that $(j, k^2 - j, k)$ is a monochromatic solution. Hence, we may assume that either $f(j) \geq j + 2k + 1$ or $f(k^2 - j) \geq k^2 - j + 2k + 1$. Coupling this with the basic $f(i) \geq i$ bound, we may now assume that

$$f(j) + f(k^2 - j) \geq k^2 + 2k + 1 = (k+1)^2$$

for all $j \in [n, n+2k]$.

Consider

$$B = \left\{ j \in [n, n+2k] : f(j) \geq \frac{(k+1)^2}{2} \right\}.$$

Let $\overline{k^2 - j}$ denote the reduction of $k^2 - j$ modulo $2k+1$ to an element of $[n, n+2k]$ and note that $f(\overline{k^2 - j}) = f(k^2 - j)$. Since we have $f(j) + f(k^2 - j) \geq (k+1)^2$ for each $j \in [n, n+2k]$ at least one of $j, \overline{k^2 - j}$ resides in B. Thus, $|B| \geq k + 1$.

Letting

$$B + B = \{x \in [n, n+2k] : x \equiv b_1 + b_2 \pmod{2k+1}; b_1, b_2 \in B\},$$

we will show that $B + B = [n, n+2k]$. Let $x \in [n, n+2k]$. Since $|(x - B)| = |B| \geq k + 1$ we cannot have $B \cap (x - B) = \emptyset$ for otherwise $|B \cup (x - B)| = |B| + |(x - B)| \geq 2k + 2$ while we know $|B + B| \leq 2k + 1$. Hence, there exist $b_1, b_2 \in B$ such that $x \equiv b_1 + b_2 \pmod{2k+1}$ has a solution for any $x \in [n, n+2k]$, and we conclude that $B + B = [n, n+2k]$.

Next, notice that for any $b \in B$ we have $f(b) \geq b + (2k+1)$ since $b \leq n + 2k$ and $k \geq 9n$ gives $f(b) \geq \frac{(k+1)^2}{2} \geq b + (2k+1)$. Hence, for all $b \in B$ we see that $g(b) \neq 0$. Thus, from Equation (2.2), we have

$$g(b) = f(b) - (2k+1) \geq \frac{(k+1)^2}{2} - (2k+1) = \frac{k^2 - 2k - 1}{2}.$$

At this point, for any $b \in B$ we have

$$b, b + (2k+1), b + 2(2k+1), \dots, g(b)$$

all of color -1 as well as the ability to find $b_1, b_2 \in B$ such that $b_1 + b_2 \equiv x \pmod{2k+1}$ for any x. Putting these together we have that, along the residue class x modulo $2k+1$, all elements between $b_1 + b_2$ and $g(b_1) + g(b_2)$ can be obtained as the sum of two elements of color -1. Note that

$$b_1 + b_2 \le 2(n + 2k) \le 2\left(\frac{k}{9} + 2k\right) \le 2\left(\frac{k}{2} + 2k\right) = 5k$$

and

$$g(b_1) + g(b_2) \ge \frac{k^2 - 2k - 1}{2} + \frac{k^2 - 2k - 1}{2} = k^2 - 2k - 1.$$

Let $m \in \left[\frac{k}{5}, \frac{4k}{5}\right] \cap \mathbb{Z}^+$ and select $b_1, b_2 \in B$ that satisfy $b_1 + b_2 \equiv m^2 \pmod{2k+1}$. We have $\frac{k^2}{25} \le m^2 \le \frac{16k^2}{25}$. Since $n \ge 14$ and $k \ge 9n$ we have $5k < \frac{k^2}{25}$. Hence,

$$\left[\frac{k^2}{25}, \frac{16k^2}{25}\right] \subseteq [5k, k^2 - 2k - 1],$$

the latter interval being the range of attainable sums of two elements of color -1. Hence, $m^2 \in [5k, k^2 - 2k - 1]$. If $\chi(m) = -1$ then we have a monochromatic solution. Otherwise, $\left[\frac{k}{5}, \frac{4k}{5}\right] \cap \mathbb{Z}^+$ consists of only elements of color 1.

Since the color pattern along any residue class modulo $2k+1$ in $[n, k^2 - n]$ is $-1, -1, \dots, -1, 1, 1, \dots, 1$ we see that $\chi(x + j(2k+1)) = 1$ for any $x \in \left[\frac{k}{5}, \frac{4k}{5}\right] \cap \mathbb{Z}^+$ and $j \in \mathbb{Z}^+$ provided $x + j(2k+1) \le k^2 - n$.

If k is even, then

$$k^2 - k = \left(\frac{k}{2} + 1\right) + \left(\frac{k}{2} - 1\right)(2k+1)$$

so we conclude that $\chi(k^2 - k) = 1$. But then $(k, k^2 - k, k)$ is a solution with all elements of color 1.

If $k \equiv 1 \pmod 4$, then both

$$\frac{k^2 - 1}{2} = \left(\frac{k-1}{4}\right) + \left(\frac{k-1}{4}\right)(2k+1)$$

and

$$\frac{k^2 + 1}{2} = \left(\frac{k+3}{4}\right) + \left(\frac{k-1}{4}\right)(2k+1)$$

have color 1, making $\left(\frac{k^2-1}{2}, \frac{k^2+1}{2}, k\right)$ a monochromatic solution.

If $k \equiv 3 \pmod 4$, then both

$$\frac{k^2 + 1}{2} + k = \left(\frac{k+1}{4}\right) + \left(\frac{k+1}{4}\right)(2k+1)$$

and

$$\frac{k^2-1}{2} - k = \binom{\frac{k+1}{4}}{} + \binom{\frac{k-3}{4}}{}(2k+1)$$

are of color 1, making $\left(\frac{k^2-1}{2} - k, \frac{k^2+1}{2} + k, k\right)$ a monochromatic solution.

Lastly, in order to conclude that $[n, (10n)^4]$ contains a monochromatic solution, note that $k \leq 80n^2$ and that the largest integer used in the proof is $k^2 - n < (80n^2)^2 - n < (10n)^4$. □

Because of Theorem 2.28, we do not have an analogue of Theorem 2.22 for nonlinear equations based only on the coefficients. This may be due to the fact that $x + y = z^2$ is not homogeneous and that with inhomogeneous polynomials we no longer have solutions preserved under dilation.

An easy modification to any backtracking van der Waerden computer program will show that every 2-coloring of $[1, 56]$ admits a monochromatic solution (other than $x = y = z = 2$) to $x + y = z^2$ (and 56 is the minimal such number). However, because solutions to $x + y = z^2$ are not preserved under dilation, this does not imply that there are infinitely many monochromatic solutions in any 2-coloring of $\mathbb{Z}^+$ as the proof of Theorem 2.28 does.

When we consider the very similar equation $x - y = z^2$, Bergelson [17] has shown this to be regular. We will not prove this here as the method of proof requires a good knowledge of either ergodic theory (see Section 5.2 for an introduction) or the algebra of the Stone-Čech compactification of the integers (see Section 5.3 for some basics).

There are a couple of parametrized families of nonlinear equations where formulas for 2-color Rado numbers are known: in [60] it is shown that $n(x-y = z^n; 2) = 2^{n+1} + 1$ and $n(ax - y = z^2; 2) = a - 1$ for any integer $a \geq 2$. Some specific numbers for various nonlinear equations for a small number of colors are given in [151].

We now delve into some more general nonlinear results. Many of the results have proofs that are well beyond the scope of this book and we will refer the reader to the appropriate references. We start with one that is easy to prove: a corollary of Rado's Theorem.

Corollary 2.29 (Multiplicative Rado's Theorem). *The equation*

$$\prod_{i=1}^{n} x_i^{c_i} = 1$$

is regular if and only if there exists $\emptyset \neq I \subseteq [1, n]$ *such that* $\sum_{i \in I} c_i = 0$.

Proof. From Theorem 2.22, we know that $\sum_{i=1}^{n} c_i y_i = 0$ is regular if and only if there exists $\emptyset \neq I \subseteq [1, n]$ such that $\sum_{i \in I} c_i = 0$. If we have a_i, $1 \leq i \leq n$, so that $\sum_{i=1}^{n} c_i a_i = 0$, then

$$2^{\sum_{i=1}^{n} c_i a_i} = \prod_{i=1}^{n} (2^{a_i})^{c_i} = 1.$$

Via the obvious bijection between colorings of $\mathbb{Z}^+$ and $2^{\mathbb{Z}^+} = \{1, 2, 4, 8, \dots\}$, the result follows. □

This corollary does not offer much new, but is included since the proof exhibits how we can translate from additive systems to multiplicative systems.

To begin our exploration of general (non-)regularity of nonlinear equations, we begin with polynomials. As noted above, the regularity of $x - y = z^2$ was originally shown by Bergelson [17] using ergodic theory. This result, as well as the non-regularity of $x + y = z^2$ (see Theorem 2.28), follow from the next, more general, result of Di Nasso and Baglini [58] who use the algebra of the Stone-Čech compactification of the integers (see Section 5.3 for basics).

Theorem 2.30. *Let $p(y)$ be a nonlinear polynomial over $\mathbb{Z}$. Then*

$$\sum_{i=1}^{n} c_i x_i = p(y)$$

is regular if and only if there exists $\emptyset \neq I \subseteq [1, n]$ such that

$$\sum_{i \in I} c_i = 0.$$

In the same paper, Di Nasso and Baglini [58] give a necessary condition for the regularity of rational functions.

Theorem 2.31. *Let $n \in \mathbb{Z}^+$. For $c_1, c_2, \dots, c_n, d_1, d_2, \dots, d_n \in \mathbb{Z} \setminus \{0\}$, if*

$$\sum_{i=1}^{n} c_i x_i^{d_i} = 0$$

is regular, then there exists $\emptyset \neq I \subseteq [1, n]$ such that

$$\sum_{i \in I} c_i = 0,$$

and, for all $i, j \in I$, we have $d_i = d_j$.

We first remark that the converse is not true. To see this, consider that $x_1^3 + x_2^3 - x_3^3 = 0$, which has $I = \{1, 3\}$ satisfying the conclusion of Theorem 2.31, has no solution in $\mathbb{Z}^+$ (Fermat's Last Theorem) and so is not even 1-regular.

As a specific example of Theorem 2.31 we see that

$$x^\ell + y^m = z^n$$

cannot be regular if $n \notin \{\ell, m\}$ (where $x = y = z = 2$ is not considered a valid monochromatic solution). In particular, $x^2 + y^2 = z$ is not regular. Of course, this does not preclude the equation we mentioned at the beginning of this section, $x^2 + y^2 = z^2$, from being regular. In fact, the regularity or non-regularity of this equation is one of the major open questions in this area.

We do know, thanks to a computing breakthrough of Heule, Kullmann, and Marek [107], that the Pythagorean triples equation is 2-regular and that

$$n(x^2 + y^2 = z^2; 2) = 7825.$$

The r-regularity of the Pythagorean triples equation for $r \geq 3$ is unknown (although it is known [133] that every r-coloring of $\mathbb{R}^+$ admits a monochromatic solution).

In a tour-de-force paper, Chow, Lindqvist, and Prendiville [40] offer the following result concerning linear combinations of equal powers.

Theorem 2.32 (Rado's Theorem for Powers). *Let $k, s \in \mathbb{Z}^+$ with $s \geq k^2 + 1$. The equation*

$$\sum_{i=1}^{s} c_i x_i^k = 0$$

is regular if and only if there exists $\emptyset \neq I \subseteq [1, s]$ such that

$$\sum_{i \in I} c_i = 0.$$

While Theorem 2.32 tells us that $v^2 + w^2 + x^2 + y^2 = z^2$ is regular, the Pythagorean triples equation's (non-)regularity determination remains elusive.

In a groundbreaking paper by Moreira [148], the following, similar, result was proved using topological dynamics (see Section 5.1 for an introduction) and the algebra of mappings.

Theorem 2.33. *Let $n \in \mathbb{Z}^+$ and assume $c_1, c_2, \ldots, c_n \in \mathbb{Z} \setminus \{0\}$. If*

$$\sum_{i=1}^{n} c_i = 0,$$

then

$$\sum_{i=1}^{n-1} c_i x_i^2 = x_n$$

is regular.

In that same paper [148], Moreira, with Theorem 2.34 below, also makes a leap forward toward proving another important open problem: Does every finite coloring of $\mathbb{Z}^+$ admit a and b such that $\{a, b, a + b, ab\}$ is monochromatic? The inclusion of both addition and multiplication makes this a significantly more difficult problem. Of course, by Schur's Theorem we can find $\{a, b, a+b\}$ monochromatic. By translating Schur's Theorem to the multiplicative setting (see the proof of Corollary 2.29) we can also find a monochromatic set $\{c, d, cd\}$. But we have no guarantee that we can take $\{a, b\} = \{c, d\}$. As a consequence of a theorem due to Hindman [110] (other results of whom we will encounter later in this book; for this one, see Theorem 2.66), it turns out that we can take the colors of $\{a, b, a+b\}$ and $\{c, d, cd\}$ to be the same.

Theorem 2.34. *The equation $x + y^2 = yz$ is regular. Consequently, every finite coloring of $\mathbb{Z}^+$ admits a and b such that $\{a, a+b, ab\}$ is monochromatic.*

The monochromatic set follows from the equation by noting that $z > y$ and taking $y = a$ and $z = a + b$, from which we must have $x = ab$.

Theorems 2.33 and 2.34 follow from a much more general result in [148], which we do not state here, with many more consequences. For example, as we have stated above, $x^2 + y^2 = z$ is not regular; however, the general result in [148] shows that $x^2 - y^2 = z$ is regular.

We now turn our attention away from polynomials. It turns out that Rado's condition on the sum of coefficients still plays a large role in the regularity of non-polynomial equations.

We start with a general result about inverses.

Theorem 2.35. *Let $\mathcal{S} = 0$ be a system of homogeneous equations. If $\mathcal{S}(x_1, x_2, \ldots, x_n) = 0$ is regular, then $\mathcal{S}\left(\frac{1}{x_1}, \frac{1}{x_2}, \ldots, \frac{1}{x_n}\right) = 0$ is regular.*

Proof. Let $r \in \mathbb{Z}^+$ and, by the Compactness Principle, let m be an integer such that every r-coloring of $[1, m]$ admits a monochromatic solution to $\mathcal{S}(x_1, x_2, \ldots, x_n) = 0$. Let $\chi : [1, m!] \to \{0, 1, \ldots, r-1\}$ be arbitrary and consider $\widehat{\chi} : [1, m] \to \{0, 1, \ldots, r-1\}$ defined by

$$\widehat{\chi}(j) = \chi\left(\frac{m!}{j}\right).$$

Let $(a_1, a_2, \ldots, a_n)$ be a solution to $\mathcal{S}(x_1, x_2, \ldots, x_n) = 0$ that is monochromatic under $\widehat{\chi}$. For $1 \leq i \leq n$, let

$$b_i = \frac{m!}{a_i}.$$

By definition of $\widehat{\chi}$ we see that $(b_1, b_2, \ldots, b_n)$ is monochromatic under χ.

Next, since $\mathcal{S}(a_1, a_2, \ldots, a_n) = 0$ and the system is homogeneous, we have

$$\mathcal{S}\left(\frac{a_1}{m!}, \frac{a_2}{m!}, \ldots, \frac{a_n}{m!}\right) = 0.$$

Hence, $\mathcal{S}\left(\frac{1}{b_1}, \frac{1}{b_2}, \ldots, \frac{1}{b_n}\right) = 0$ is a monochromatic (under χ) solution. Since χ is arbitrary, we are done. □

In order to move to more general non-polynomials, consider Lemma 2.37 below, which uses the following definition.

Definition 2.36. Let $(R, +, \cdot)$ be a ring and let $A \subseteq R \setminus \{0\}$. We say that a system $\mathcal{S}$ of equations from $R[X_1, X_2, \ldots, X_n]$ is *regular over* A if every finite coloring of A admits a monochromatic solution to $\mathcal{S}$.

Lemma 2.37. *Let* $(R, +, \cdot)$ *be a ring. For* $A, B \subseteq R \setminus \{0\}$, *let* $f : A \to B$ *be a bijection. Let* $\mathcal{S} = 0$ *be a system of equations with coefficients in* R. *Then* $\mathcal{S}(x_1, x_2, \ldots, x_n) = 0$ *is regular over* A *if and only if* $\mathcal{S}(f^{-1}(x_1), f^{-1}(x_2), \ldots, f^{-1}(x_n)) = 0$ *is regular over* B.

Proof. Since f^{-1} is also a bijection, it suffices to prove only one direction. To this end, assume that $\mathcal{S}(x_1, x_2, \ldots, x_n) = 0$ is regular over A. For $r \in \mathbb{Z}^+$, let $\chi : B \to \{0, 1, \ldots, r-1\}$ be arbitrary and consider $\widehat{\chi} : A \to \{0, 1, \ldots, r-1\}$ defined by $\widehat{\chi}(a) = \chi(f(a))$. Let $(a_1, a_2, \ldots, a_n)$ be a solution that is monochromatic under $\widehat{\chi}$.

Set $b_i = f(a_i)$ for $1 \leq i \leq n$ and notice that, from the definition of $\widehat{\chi}$, we have all b_i of the same color under χ. Furthermore, we see that $\mathcal{S}\big(f^{-1}(b_1), f^{-1}(b_2), \ldots, f^{-1}(b_n)\big) = 0$ so that $\mathcal{S}\big(f^{-1}(x_1), f^{-1}(x_2), \ldots, f^{-1}(x_n)\big)$ is r-regular. Since χ and r are arbitrary, we are done. □

Theorem 2.38 (Lefmann's Theorem)**.** *Let* $n \in \mathbb{Z}^+$ *and* $c_1, c_2, \ldots, c_n \in \mathbb{Z}\setminus\{0\}$. *For* $k \in \mathbb{Z} \setminus \{0\}$, *the equation*

$$\sum_{i=1}^{n} c_i x_i^{\frac{1}{k}} = 0$$

is regular if and only if there exists $\emptyset \neq I \subseteq [1, n]$ *such that*

$$\sum_{i \in I} c_i = 0.$$

Proof. First, consider $k \in \mathbb{Z}^+$. By Rado's Theorem, we know that the result holds for $k = 1$. Applying Lemma 2.37 to the ring $(\mathbb{Z}, +, \cdot)$ with $A = \mathbb{Z}^+$,

$$B = \{i^k : i \in \mathbb{Z}^+\},$$

and $f : A \to B$ defined by $f(a) = a^k$ proves the theorem for all positive integers k.

For $k \in \mathbb{Z}^-$, we may now appeal to Theorem 2.35. □

Remark. Theorem 2.35, due to Brown and Rödl [30], contains the case $k = -1$ of Theorem 2.38 (due to Lefmann [133]) and was independently proved.

Corollary 2.39. *Let* $r \in \mathbb{Z}^+$. *Every* r*-coloring of* $\mathbb{Z}^+$ *admits a monochromatic solution to*

$$\sum_{i=1}^{m} \frac{1}{x_i} = 1,$$

for some $m \geq 3$.

Proof. Apply Theorem 2.38 (or Theorem 2.35) to $\frac{1}{y} = \sum_{i=1}^{n} \frac{1}{x_i}$ and note that

$$1 = \sum_{i=1}^{n} \Big(\underbrace{\frac{1}{x_i} + \frac{1}{x_i} + \cdots + \frac{1}{x_i}}_{y \text{ times}} \Big).$$

Taking $m = yn$ finishes the proof. □

In the proof of Corollary 2.39 we see that allowing the variables to appear repeatedly is essential. Can we prove the same result if we require all x_i to be distinct? This is a much harder question that was proved by Croot in [52]. The proof of Croot's Theorem is analytic number theoretic in nature and is beyond the scope of this book.

Theorem 2.40 (Croot's Theorem). *Every r-coloring of $\mathbb{Z}^+$ admits a monochromatic set $S \neq \{1\}$ such that*

$$\sum_{s \in S} \frac{1}{s} = 1.$$

Remark. Croot actually shows that every r-coloring of $[2, e^{167000r}]$ suffices.

The last equation we will explore in this section incorporates exponentiation. As we've seen, we can guarantee monochromatic sets $\{a, b, a+b\}$ and $\{c, d, cd\}$. The next logical set in this direction would be $\{x, y, x^y\}$, i.e., solutions to $z = x^y$. In 2011, Sisto [190] proved that this equation is 2-regular. Five years later, Sahasrabudhe [177] proved that the equation is regular. In fact, more was shown:

Theorem 2.41. *Every finite coloring of $\mathbb{Z}^+$ contains a monochromatic solution to the system $z = x^y, w = xy$, with $x, y > 1$.*

Hence, we now see that we can guarantee a monochromatic set of the form $\{a, b, ab, a^b\}$.

The proof of Theorem 2.41 is quite lengthy and technical. We will prove only that $z = x^y$ is 2-regular by following the proof given by Sisto [190].

Proof of the 2-regularity of $z = x^y$. By translating van der Waerden's Theorem to the multiplicative setting, we find that every 2-coloring of $\mathbb{Z}^+$ admits arbitrarily long monochromatic geometric progressions.

Consider an arbitrary 2-coloring of $\mathbb{Z}^+$ defined by the partition $\mathbb{Z}^+ = A \sqcup B$. Without loss of generality, A must contain arbitrarily long geometric progressions.

Define the sets

$$A_n = \{x \in A : n^x \in A\} \quad \text{and} \quad B_n = \{x \in A : n^x \in B\}.$$

If for all $n \in \mathbb{Z}^+ \setminus \{1\}$ we have $A_n \neq \emptyset$ then for any $a \in A$ there exists

$b \in A_a$ so that $a^b \in A$. Hence, we have $\{a, b, a^b\} \subseteq A$ and are done in this circumstance.

So, we may assume that there exists $m \in \mathbb{Z}^+ \setminus \{1\}$ such that $A_m = \emptyset$ so that $B_m = A$.

Let k be the minimal element in A. Since A contains arbitrary long geometric progressions, we take one of length $k+1$: we take $a, s \in \mathbb{Z}^+$ such that $a, as, as^2, \ldots, as^k$ all reside in B_m.

If $s^i \in B$ for some $i \in \{1, 2 \ldots, k\}$ then

$$\left\{m^a, s^i, m^{as^i}\right\} \subseteq B$$

and we are done. Otherwise, $s^i \in A$ for all $i \in \{1, 2, \ldots, k\}$, meaning $\{s, k, s^k\} \subseteq A$ is the desired monochromatic set. □

2.2.4 Algebraic Equations

Clearly $\mathbb{Z}^+$ is closed under addition and multiplication; however, it does not contain nontrivial multiplicative or additive inverses. This is one reason that the (non-)regularity over $\mathbb{Z}^+$ of some of the equations in the last subsection is hard to prove. If we consider fields, we will have more tools at our disposal and may be able to prove more. This does not mean the results are easy, just that we are able to bring in more algebra and number theory. We will restrict our attention to prime order fields, i.e., $\mathbb{F}_p$ with p prime.

A word of caution: positive results over fields do not necessarily imply that the corresponding result over $\mathbb{Z}^+$ is true. For example, if we consider $x^3 + y^3 = z^3$, we know that there is no solution over $\mathbb{Z}^+$, while it is easy to find solutions over prime-order fields $\mathbb{F}_p$, e.g., $3^3 + 5^3 \equiv 4^3 \pmod{11}$. Note, however, if we can find that an equation is not r-regular over any sufficiently large prime-order fields, then we can conclude the non-regularity over $\mathbb{Z}^+$ via the contrapositive of the Compactness Principle.

With regards to the Fermat equations, it is known [186] that for any $k \in \mathbb{Z}^+$, for all sufficiently large primes p, the equation $x^k + y^k = z^k$ is regular over $\mathbb{F}_p$, i.e., $x^k + y^k \equiv z^k \pmod{p}$ has a monochromatic solution with $xyz \not\equiv 0 \pmod{p}$ for any r-coloring of $\mathbb{F}_p$ for p a sufficiently large prime. We will not prove this result here; it is shown in Section 7.1.

All results in this subsection will be presented without proof as they use methods not commonly encountered in undergraduate mathematics. We will, however, in Sections 2.5, 5.2, and 5.4, present introductions to some of the tools used in these proofs and apply these tools to other, more fundamental, results.

Theorem 2.42. *For any $r \in \mathbb{Z}^+$, there exists a prime $P(r)$ such that the equation $x + y = uv$ is r-regular over $\mathbb{F}_p$ for all primes $p \geq P(r)$.*

The above theorem is due to Sárközy [181]. After reading through Section

2.5, the interested reader should be able to work through its proof. Theorem 2.42 was extended to prime power fields in [53].

As shown in the last section, $x + y = z^2$ is not regular – in fact, it is not even 3-regular – over $\mathbb{Z}^+$. The situation over $\mathbb{F}_p$ is much different.

Theorem 2.43. *For any $r \in \mathbb{Z}^+$, there exists a prime $P(r)$ such that the equation $x + y = z^2$ is r-regular over $\mathbb{F}_p$ for all primes $p \geq P(r)$.*

Theorem 2.43 is due to Lindqvist [135], who shows more generally that $x^j + y^k = z^\ell$ is regular over $\mathbb{F}_p$ (compare this with Theorem 2.31 and its subsequent discussion).

Lindqvist, in [135], uses techniques employed by Green and Sanders in [89] to prove the following theorem, the finite field analogue of the $\{a, b, a + b, ab\}$ regularity (over $\mathbb{Z}^+$) open question.

Theorem 2.44. *For any $r \in \mathbb{Z}^+$, there exists a prime $P(r)$ such that the system $z = x + y, w = xy$ is r-regular over $\mathbb{F}_p$ for all primes $p \geq P(r)$. Consequently, every r-coloring of $\mathbb{F}_p$ admits a monochromatic set of the form $\{a, b, a + b, ab\}$ for all sufficiently large primes p.*

2.3 Hales-Jewett Theorem

We now come to a fundamental Ramsey object. In 1963, Hales and Jewett [96] produced a new proof of van der Waerden's Theorem by discovering a more general – and significantly stronger – result, from which van der Waerden's Theorem follows readily. Indeed, Deuber's Theorem (on (m, p, c)-sets; see Theorem 2.13) can be deduced from their result (although not easily); see [131] and also Prömel's proof as reproduced in [93].

As the Hales-Jewett Theorem is a multi-dimensional theorem, we have need of the following standard notation.

Notation. For any set S, we let $S^n = \underbrace{S \times S \times \cdots \times S}_{n \text{ copies}}$.

In order to state Hales and Jewett's result, we use the following definition. (We will use the language of variable words; other authors refer to evaluated variable words as combinatorial lines, but we will not use this language.)

Definition 2.45 (Variable word). A word of length m over the alphabet $\mathcal{A}$ is an element of $\mathcal{A}^m$, which we may write as $a_1 a_2 \ldots a_m$ with $a_i \in \mathcal{A}$ for all $1 \leq i \leq m$. Let x be a variable which may take on any value in $\mathcal{A}$. A word $w(x)$ over the alphabet $\mathcal{A} \cup \{x\}$ is called a *variable word* if x occurs in $w(x)$. For $i \in \mathcal{A}$, the word $w(i)$ is obtained by replacing each occurrence of x with i.

Theorem 2.46 (Hales-Jewett Theorem). *Let $k, r \in \mathbb{Z}^+$. Let $\mathcal{W}(m)$ be the set of all variable words of length m over the alphabet $\{1, 2, \ldots, k\} \cup \{x\}$. There exists a minimal positive integer $H = HJ(k; r)$ such that for any $h \geq H$, every r-coloring of the elements of $[1, k]^h$ admits $w(x) \in \mathcal{W}(h)$ with $\{w(i) : i \in [1, k]\}$ monochromatic.*

As should be clear by now, the numbers $HJ(k; r)$ are referred to as *Hales-Jewett numbers.* Trivially, we have $HJ(k; 1) = 1$ since taking the only variable word of length 1, i.e., $w(x) = x$, we see that $\{w(i) : i \in [1, k]\}$ is monochromatic since we are using only one color.

In our discussion below, under a given coloring, if a variable word $w(x)$ results in $\{w(i) : i \in [1, k]\}$ being monochromatic, we will say we have a *monochromatic variable word.*

Before delving into the proof, it will be useful to consider a few examples to help understand this theorem.

Example 2.47. We will show that $HJ(3; 2) > 2$. We will show this by considering the children's game Tic-Tac-Toe. As we know, Tic-Tac-Toe usually ends in a tie. Consider the end result of a tied game in Figure 2.1.

O	X	X
X	X	O
O	O	X

FIGURE 2.1
Tic-Tac-Toe ends in a draw, showing $HJ(3; 2) > 2$

We first translate this to a 2-coloring of $[1, 3]^2$ via matrix notation, i.e., the Xs and Os are colors of the entries $(a_{i,j})_{1 \leq i \leq 3}$. So, each of $a_{1,1}$, $a_{2,3}$, $a_{3,1}$, and $a_{3,2}$ has color O while the color of all others is X. By using only the subscripts, we see that we have a 2-coloring of the elements of $[1, 3]^2$.

If we had $HJ(3; 2) = 2$ then we would be guaranteed a monochromatic variable word $w(x)$ over the alphabet $[1, 3] \cup x$ of length 2. So, let's consider all possible variable words.

Let $W = \{w(x) : x \in \{1, 2, 3\}\}$. If $w(x) = x\,x$ then $W = \{11, 22, 33\}$ which corresponds to the main diagonal of our Tic-Tac-Toe board. If $w(x) = ax$ for some $a \in \{1, 2, 3\}$, then $W = \{a1, a2, a3\}$, which is the ath row of the Tic-Tac-Toe board. If $w(x) = xa$, then W represents the ath column. (Note that we do not have the antidiagonal defined by any variable word.)

As you can see from the Tic-Tac-Toe end result in Figure 2.1, there is no winner. This means there is no monochromatic variable word so we can deduce that $HJ(3; 2) > 2$.

Remark. Hales and Jewett's motivation for [96] was a multi-dimensional Tic-Tac-Toe game analysis played on $[1, k]^n$. They proved, using Theorem 2.46, that the first player can force a win provided $n = n(k)$ is sufficiently large (there are two colors, X and O, one for each player).

Example 2.48. We will show that $HJ(2;r) = r$. To see that $HJ(2;r) \le r$ consider the following $r+1$ words of length r over the alphabet $\{1,2\}$:

$$\begin{array}{c} 11\cdots 111 \\ 11\cdots 112 \\ 11\cdots 122 \\ \vdots \\ 12\cdots 222 \\ 22\cdots 222. \end{array}$$

Since we are using r colors, two of these words, say u and v, have the same color. Hence, for some variable word $w(x)$ of length r of the form $1\cdots 1x\cdots x2\cdots 2$, we have $w(1) = u$ and $w(2) = v$. As u and v have the same color, we have shown that words of length r suffice. To see that $HJ(2;r) \ge r$ we will find an r-coloring of all words of length $r-1$ over $\{1,2\}$ that does not admit a monochromatic variable word. Using the colors $\{0,1,\dots,r-1\}$, define an r-coloring of the length $r-1$ words by coloring $w_1w_2\dots w_{r-1}$ with color

$$\sum_{i=1}^{r-1} w_i - (r-1).$$

For any variable word $w(x)$, we see that the colors of $w(1)$ and $w(2)$ must be different, giving us $HJ(2;r) > r-1$.

Let's dive into some preliminaries needed in the proof of the Hales-Jewett Theorem. We start with a lemma that requires the following definitions.

Definition 2.49 (Prefix)**.** Let $\mathcal{A}$ be an alphabet. Let $m,n \in \mathbb{Z}^+$. If u is a word over $\mathcal{A}$ of length m and v is a word over the same alphabet of length n, then the concatenation $uv = u_1u_2\dots u_mv_1v_2\cdots v_n$ is a word over $\mathcal{A}$ of length $m+n$. We call u a *prefix* of v.

The next definition gives us a means for translating a coloring of $[1,k]^{m+n}$ to a coloring of $[1,k]^n$.

Definition 2.50 (Inflated Coloring)**.** Let $k,m,n,r \in \mathbb{Z}^+$ and let χ be an r-coloring of the words in $[1,k]^{m+n}$. For any given word of length $m+n$, we have uv with u a prefix of length m and v a word of length n. For a fixed $v \in [1,k]^n$, consider all prefixes of length m. The *inflated coloring* of χ for $[1,k]^n$, denoted $\dot{\chi}$, is the r^{k^m}-coloring of $[1,k]^n$ defined by

$$\dot{\chi}(v) = (\chi(u_1v), \chi(u_2v), \dots, \chi(u_{k^m}v))$$

for any predetermined ordering $u_1 \prec u_2 \prec \cdots \prec u_{k^m}$ of all possible prefixes of length m.

Definition 2.50 allows us to shorten the length of words at the expense of the number of colors. This will help with the induction argument we use in the next lemma.

Lemma 2.51. *Let $\ell \in \mathbb{Z}^+$ be fixed and assume that $HJ(\ell; s)$ exists for all $s \in \mathbb{Z}^+$. For a fixed $r \in \mathbb{Z}^+$, let $n_0 = HJ(\ell; r)$ and, for $i \in \mathbb{Z}^+$, define*

$$n_i = HJ(\ell; r^{\ell^{(n_0+n_1+\cdots+n_{i-1})}}).$$

Then, for any nonnegative integer j, any r-coloring of

$$[1,\ell]^{(n_0+n_1+\cdots+n_j)} \cong [1,\ell]^{n_0} \times [1,\ell]^{n_1} \times \cdots \times [1,\ell]^{n_j}$$

admits variable words $w_i(x)$ in $[1,\ell]^{n_i}$, $0 \le i \le j$, such that χ is constant on

$$\{w_0(i) : i \in [1,\ell]\} \times \{w_1(i) : i \in [1,\ell]\} \times \cdots \times \{w_j(i) : i \in [1,\ell]\}.$$

Proof. Let χ be an arbitrary r-coloring of $[1,\ell]^{(n_0+n_1+\cdots+n_j)}$. Writing

$$[1,\ell]^{(n_0+n_1+\cdots+n_j)} = [1,\ell]^{(n_0+n_1+\cdots+n_{j-1})} \times [1,\ell]^{n_j},$$

consider the inflated coloring $\dot{\chi}$ of χ for $[1,\ell]^{n_j}$, which uses (at most) $r^{\ell^{(n_0+n_1+\cdots+n_{j-1})}}$ colors. By definition of n_j we have, under $\dot{\chi}$, a monochromatic variable word $w_j(x)$ of length n_j. With a slight abuse of notation, this means that if u is any prefix of $w_j(x)$ of length $n_0 + n_1 + \cdots + n_{j-1}$, then for all $a, b \in [1,\ell]$ we have

$$\chi(uw_j(a)) = \chi(uw_j(b)).$$

By viewing χ restricted to

$$[1,\ell]^{(n_0+n_1+\cdots+n_{j-1})} \times \{w_j(i) : i \in [1,\ell]\},$$

we now have a way to restrict χ to $[1,\ell]^{(n_0+n_1+\cdots+n_{j-1})}$ in a well-defined manner.

We repeat this process until we obtain

$$[1,\ell]^{n_0} \times \{w_1(i) : i \in [1,\ell]\} \times \cdots \times \{w_j(i) : i \in [1,\ell]\}$$

with the property that

$$\chi(uw_1(i_1)w_2(i_2)\ldots w_j(i_j)),$$

for any $1 \le i_1, i_2, \ldots, i_j \le \ell$, depends only on the prefix $u \in [1,\ell]^{n_0}$. Hence, we can now view χ as an r-coloring of $[1,\ell]^{n_0}$. By definition of n_0, we have a monochromatic variable word $w_0(x)$ in $[1,\ell]^{n_0}$, thereby finishing the proof. □

The usefulness of Lemma 2.51 is that we have several monochromatic variable words, which affords us the flexibility to prove the Hales-Jewett Theorem.

Proof of Theorem 2.46. We use double induction on k and r. We have already shown that the theorem holds for $(k;r) \in \{(k;1),(2;r)\}$. Hence, we assume that $h = HJ(k;r-1)$ exists and that $HJ(k-1;s)$ exists for all s. We will show that $HJ(k;r)$ exists.

Call a variable word $v(x)$ over $[1,k] \cup \{x\}$ *almost-monochromatic* if $\{v(i) : i \in [1,k-1]\}$ is monochromatic.

Consider

$$n = \sum_{i=0}^{h} n_i,$$

where the n_i are as defined in Lemma 2.51 with $\ell = k-1$. We will show that $HJ(k;r) \leq n$.

Consider an arbitrary r-coloring χ of $[1,k]^n$. From the assumption that $h = HJ(k-1;r)$ exists, Lemma 2.51 (with $j = h$) gives us variable words $w_i(x)$ of length n_i, for $0 \leq i \leq h$, such that

$$w_0(x_1) \times w_1(x_2) \times \cdots \times w_h(x_h)$$

is almost-monochromatic, say of color c.

Consider

$$S = \{w_0(k)\} \times w_1(x_1) \times w_2(x_2) \times \cdots \times w_h(x_h),$$

and for $(a_1, a_2, \ldots, a_h) \in [1,k]^h$ let

$$S(a_1, a_2, \ldots, a_h) = \{w_0(k)\} \times \{w_1(a_1)\} \times \{w_2(a_2)\} \times \cdots \times \{w_h(a_h)\}.$$

We view $S(a_1, a_2, \ldots, a_h)$ as a word of length n, so that it is assigned a color under χ. Let

$$\mathfrak{S} = \bigcup_{(a_1,\ldots,a_h)\in[1,k]^h} S(a_1, a_2, \ldots, a_h).$$

If $\chi(S(a_1, a_2, \ldots, a_h)) \neq c$ for all $(a_1, a_2, \ldots, a_h) \in [1,k]^h$ then we have an $(r-1)$-coloring of $\mathfrak{S}$. Using the coloring of $\mathfrak{S}$, define the $(r-1)$-coloring γ of $[1,k]^h$ by

$$\gamma(a_1 a_2 \ldots a_h) = \chi(S(a_1, a_2, \ldots, a_h)).$$

Since $h = HJ(k;r-1)$ exists by the inductive assumption, we see that γ admits a monochromatic variable word

$$v(x) = v_1 v_2 \ldots v_h$$

over $[1,k] \cup \{x\}$, where $v_i = x$ for at least one $i \in \{1,2,\ldots,h\}$. This means that (with a slight abuse of notation)

$$\{w_0(k)\} \times \{w_1(v_1)\} \times \{w_2(v_2)\} \times \cdots \times \{w_h(v_h)\}$$

is a monochromatic variable word under χ.

Hence, we may assume there exists $(b_1, b_2, \ldots, b_h) \in [1, k]^h$ with

$$\chi(S(b_1, b_2, \ldots, b_h)) = c.$$

For each occurrence of $b_i = k$ let $b_i = x$ and consider (again, with a slight abuse of notation)

$$u(x) = w_0(x) \times \{w_1(b_1)\} \times \{w_2(b_2)\} \times \cdots \times \{w_h(b_h)\}.$$

Notice that the first coordinate contains the variable x so that this is indeed a variable word over the alphabet $[1, k-1] \cup \{x\}$.

By the inductive assumption, $u(x)$ is almost-monochromatic of color c, while by construction $u(k)$ also has color c. Hence, $u(x)$ is a monochromatic variable word over $[1, k] \cup \{x\}$, thereby proving the theorem. □

As promised, this theorem implies van der Waerden's Theorem. We present an argument (noted in [13]) due to Shelah [189] as it gives a better lower bound on $HJ(k; r)$ (see Theorem 2.52, below) than the standard argument which associates $[1, k]^n$ with $[1, k^n]$.

Let $n = HJ(k; r)$ and consider an arbitrary r-coloring χ of $[1, n(k-1)+1]$. We will show that χ admits a monochromatic k-term arithmetic progression.

Consider $[1, k]^n$ and for each word $a_1 a_2 \ldots a_n$ associate the integer

$$1 + \sum_{i=1}^{n} (a_1 - 1).$$

We color every word by the color of its associated integer under χ.

By the Hales-Jewett Theorem we have a monochromatic variable word $w(x)$ of length n. Viewing this as a "variable integer" we see that $w(x)$ is

$$a + (x-1)|I|$$

for some non-empty $I \subseteq \{1, 2, \ldots, n\}$, where

$$a = 1 + \sum_{i \in [1,n] \setminus I} (a_i - 1).$$

Thus $w(1), w(2), w(3), \ldots, w(k)$ is associated with $a, a+d, a+2d, \ldots, a+(k-1)d$ where $d = |I|$. Since $w(x)$ is monochromatic, we have our monochromatic k-term arithmetic progression.

The above argument that the Hales-Jewett Theorem implies van der Waerden's Theorem gives the following lower bound on $HJ(k; r)$.

Theorem 2.52. *For $k, r \in \mathbb{Z}^+$ we have*

$$HJ(k; r) \geq \frac{w(k; r) - 1}{k - 1}.$$

Using the best-known lower bound for $w(k;r)$ (found in [126]) we have the following immediate corollary.

Corollary 2.53. *There exists a constant $c > 0$ such that, for all positive integers k and r we have*

$$HJ(k;r) \geq c \cdot \frac{r^{k-1}}{k-1}.$$

The first nontrivial Hales-Jewett number was recently determined in [112] to be $HJ(3;2) = 4$. In Example 2.47 we showed that $HJ(3;2) \geq 3$. To see that $HJ(3;2) > 3$ consider the 2-coloring of a $3 \times 3 \times 3$ cube in Figure 2.2 and view it as a 3-dimensional Tic-Tac-Toe game. The $\star$ label designates the $(1,1,1)$ position while the $\bullet$ label signifies the $(3,1,2)$ position.

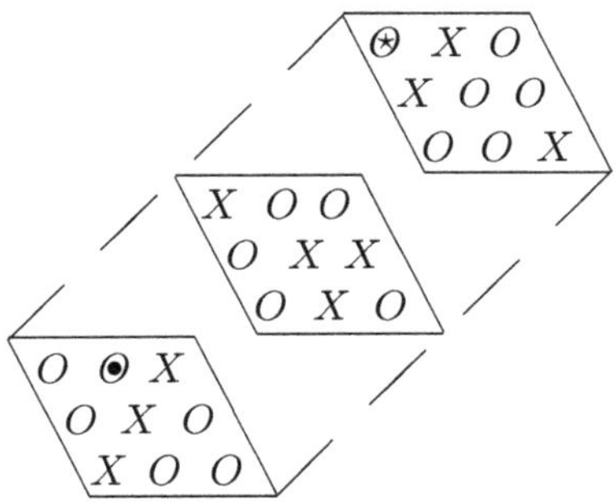

FIGURE 2.2
Three-dimensional Tic-Tac-Toe giving $HJ(3;2) > 3$

Note that, although the antidiagonal on the top board is a winning Tic-Tac-Toe position, this is not represented by any monochromatic variable word. There are other monochromatic antidiagonals, too (this is to be expected, since the $3 \times 3 \times 3$ Tic-Tac-Toe game always has a winner). However, these antidiagonals are the only winning 3-in-a-row configurations in Figure 2.2. Hence, no variable word is monochromatic.

Although the parameters we are discussing are small, the search spaces are quite large due to the multi-dimensional nature of Hales-Jewett numbers and, at this point, require extensive computer time.

2.4 Finite Sums

We have encountered finite sums in Section 2.1.1 when we considered Hilbert's Cube Lemma. The definition we present below is more standard for finite sums than that which we presented in Definition 2.9 (where we allowed repeated elements).

Definition 2.54 (Finite sums, sumset). Let $S = \{s_i\}_{i=1}^{\infty} \subseteq \mathbb{Z}^+$. The set of *finite sums* of S, denoted $FS(S)$, is defined by

$$FS(S) = \left\{ \sum_{i \in I} s_i : \emptyset \neq I \subseteq \mathbb{Z}^+, \, |I| < \infty \right\}.$$

We also refer to $FS(S)$ as a *sumset*.

We will start by considering the situation when S is finite.

2.4.1 Arnautov-Folkman-Rado-Sanders' Theorem

The result named in this subsection is a mouthful. This stems from the independent, and distinct, proofs used to prove the result that all occurred around the same time. From historical research done by Soifer [191] we find that Arnautov in [9], Folkman via personal communication, and Sanders in [178], all proved the same result, which is a generalization of Schur's Theorem. As noted in [84], the result also follows directly from Rado's Theorem (hence the inclusion of Rado's name). We will offer two different proofs in this subsection: one based on van der Waerden's Theorem (due to Folkman) and one based on Rado's Theorem. A third proof will be given in Section 3.3.2 as it appeals to a result presented in the next chapter. Other (different) proofs can be found in [152, 199].

Theorem 2.55 (Arnautov-Folkman-Rado-Sanders' Theorem). *Let $k, r \in \mathbb{Z}^+$. There exists a minimal positive integer $n = n(k;r)$ such that every r-coloring of $[1,n]$ admits $S \subseteq [1,n]$ with $|S| = k$ such that $FS(S)$ is monochromatic.*

Proof based on van der Waerden's Theorem. We start by defining the auxiliary function $a = a(k;r)$ as the minimal positive integer such that every r-coloring of $[1,a]$ admits $B = \{b_1 < b_2 < \cdots < b_k\}$ with $FS(B) \subseteq [1,a]$ such that

$$\chi\left(b_{i_1} + b_{i_2} + \cdots + b_{i_j}\right) = \chi(b_{i_j})$$

for any $j \in [1,k]$ with $i_1 < i_2 < \cdots < i_j$.

Assume for the moment that $a(k;r)$ exists for all $k, r \in \mathbb{Z}^+$. Then, by considering $a(r(k-1)+1;r)$, we have $B = \{b_1 < b_2 < \cdots < b_{r(k-1)+1}\}$. By the pigeonhole principle, there must be at least k elements of B of the same color. Say $S = \{b_{i_1}, b_{i_2}, \cdots, b_{i_k}\}$ is monochromatic. Since $FS(S) \subseteq FS(B)$ we see that every element of $FS(S)$ is colored by the largest element of the sum. Since this largest element is necessarily in S, and since S is monochromatic, we see that $FS(S)$ is monochromatic.

So, we will be done once we establish the existence of $a(k;r)$. We proceed by induction on k, with $k = 1$ being trivial. Hence, let $\ell = a(k-1;r)$ and let

$m = w(\ell + 1; r)$. We will show that $a(k;r) \leq 2m$. To this end, consider any r-coloring of $[1, 2m]$. By van der Waerden's Theorem, there exist

$$c, c+d, \ldots, c+\ell d \in [m+1, 2m]$$

all the same color. Note that $c > m$.

Since sumsets are closed under dilation, we can apply the inductive assumption to $d[1, \ell]$ to obtain

$$B = \{db_1 < db_2 < \cdots < db_{k-1}\}$$

with the desired property. Note that $d[1, \ell] \subseteq [1, m]$ since $c + \ell d \leq 2m$ and $c > m$. Let $b_k = c$ and notice that $db_{k-1} < b_k$. As every element of $FS(B)$ is a multiple of d, with largest possible multiple $d\ell$ (the sum of all elements of B must remain in $d[1, \ell]$ by definition of $a(k-1;r)$; otherwise elements larger than $a(k-1;r)$ would not be assigned a color). Hence, if $s \in FS(B)$ we see that $s + b_k = c + jd$ for some $j \in [1, \ell]$. Since all members of $c+d, \ldots, c+\ell d$ are the same color as $c = b_k$, we have $a(k;r) \leq 2m$. □

Proof based on Rado's Theorem. Consider the following system of linear homogeneous equations:

$$\left\{ \sum_{i \in I} x_i = y_I : \emptyset \neq I \subseteq [1, k] \right\},$$

where we index the y's by the subset over which we are summing. By showing a monochromatic solution to this system exists we will have values of the variables for which $FS(x_1, x_2, \ldots, x_k)$ is monochromatic. By Rado's Theorem we will be done once we show that the associated coefficient matrix satisfies the columns condition.

This follows quickly by induction on k. For $k = 2$ the system is

$$\begin{pmatrix} 1 & 0 & -1 & 0 & 0 \\ 0 & 1 & 0 & -1 & 0 \\ 1 & 1 & 0 & 0 & -1 \end{pmatrix} \begin{pmatrix} x_1 \\ x_2 \\ y_{\{1\}} \\ y_{\{2\}} \\ y_{\{1,2\}} \end{pmatrix} = \mathbf{0}.$$

Letting $\mathbf{c}_1, \mathbf{c}_2, \ldots, \mathbf{c}_5$ be the columns in the coefficient matrix, from left to right, we have $\mathbf{c}_1 + \mathbf{c}_3 + \mathbf{c}_5 = \mathbf{0}$ and $\mathbf{c}_2 + \mathbf{c}_4 = -\mathbf{c}_5$. Hence, for $k = 2$ the columns condition is satisfied.

We now assume the result for $k-1$ and will show it for k. The coefficient matrix has form (where the top row is for reference and is not part of the matrix):

$$\left(\begin{array}{ccccc} x_1 \to x_{k-1} & x_k & y_{\{1\}} \to y_{[1,k-1]} & y_{\{k\}} & y_{\{1,k\}} \to y_{[1,k]} \\ & 0 & & 0 & \\ B & \vdots & -I & \vdots & \mathcal{O} \\ & 0 & & 0 & \\ \hline 0 \ \cdots \ 0 & 1 & & -1 & 0 \ \cdots \ 0 \\ & 1 & & 0 & \\ B & \vdots & \mathcal{O} & \vdots & -I \\ & 1 & & 0 & \end{array}\right),$$

where I is the identity matrix, $\mathcal{O}$ is the matrix of all 0s, and B is the matrix with 1s precisely in the columns given in the index of y_I for which -1 is in the y_I column. The leftmost set of y_I's represent all subsets that do not include k; the second set are all those subsets that do include k.

Note that B is the same in both positions by the obvious bijection between subsets including k and subsets not including k. Furthermore, this bijection gives us that I and $\mathcal{O}$ have the same dimensions.

Note that the sum of the x_k and $y_{\{k\}}$ columns can be written as a linear combination of the columns $y_{\{1,k\}} \to y_{[1,k]}$:

$$x_k + y_{\{k\}} = - \sum_{\substack{I \subseteq [1,k] \\ k \in I}} y_I.$$

So, consider the matrix with the x_k and $y_{\{k\}}$ columns deleted:

$$\begin{pmatrix} B & -I & \mathcal{O} \\ 0 \cdots 0 & 0 \cdots 0 & 0 \cdots 0 \\ B & \mathcal{O} & -I \end{pmatrix}.$$

By the inductive assumption we see that the matrix $(B \quad -I \quad \mathcal{O})$ satisfies the columns condition and, hence, so does the above matrix. Since $x_k + y_{\{k\}}$ is a linear combination of the other columns, we see that our original matrix satisfies the columns condition, and we are done. □

Remark. Both Schur's Theorem and Hilbert's Cube Lemma are direct consequences of the Arnautov-Folkman-Rado-Sanders Theorem.

Although Theorem 2.55 has a cumbersome name, we will not have need to refer to it since it is subsumed by Hindman's more powerful theorem, which we visit next.

2.4.2 Hindman's Theorem

We are still investigating finite sums $FS(S)$; however, we will now be considering the situation when S is infinite. In particular, does the conclusion of Theorem 2.55 hold when S is infinite? As you may suspect if you read the title of this subsection, an answer was achieved by Hindman [109].

Theorem 2.56 (Hindman's Theorem). *Let $r \in \mathbb{Z}^+$. Every r-coloring of $\mathbb{Z}^+$ admits an infinite set $A \subseteq \mathbb{Z}^+$ such that $FS(A)$ is monochromatic.*

The proof of this must be fundamentally different than the proof when S is finite. We can use neither van der Waerden's Theorem nor Rado's Theorem as these will only give us arbitrarily large – not infinite – sets, since both are results about finite structures.

The proof of Hindman's Theorem is quite involved, so we will break it up into several lemmas before putting the pieces together. We will follow the proof given by Baumgartner [11]. But first, a definition is in order.

Definition 2.57 (Disjoint sumset of S). Let $S, T \subseteq \mathbb{Z}^+$ be infinite sets. We say $T \subseteq FS(S)$ is a *disjoint sumset of* S, and write $T \in \mathcal{DS}(S)$, if every element of S is contained in at most one sum/element in T, where $\mathcal{DS}(S)$ is the class of all disjoint sumsets of S.

The benefit of this class of finite sums is that if $t_1, t_2 \in T \in \mathcal{DS}(S)$ then $t_1 + t_2 \in FS(S)$ since the elements from S used in t_1 are distinct from those used in t_2. Using this idea, we immediately have the following lemma, the proof of which is left to the reader as Exercise 2.17.

Lemma 2.58. *Let $S, T, U \subseteq \mathbb{Z}^+$ be infinite sets with $T \in \mathcal{DS}(S)$ and $U \in \mathcal{DS}(T)$. Then*

(i) $FS(T) \subseteq FS(S)$; and

(ii) $U \in \mathcal{DS}(S)$.

Before the next lemma, we require another definition.

Definition 2.59 (Intersective for S). Let $S \subseteq \mathbb{Z}^+$ be infinite. We say that a set X is *intersective for* S if for all $T \in \mathcal{DS}(S)$ we have $FS(T) \cap X \neq \emptyset$.

A crucial observation here is that any intersective set must be infinite. This can be confirmed by part (ii) of the next lemma.

Lemma 2.60. *Let X, S be subsets of $\mathbb{Z}^+$. The following hold:*

(i) Let $n \in \mathbb{Z}^+$ and assume $X = \bigsqcup_{i=1}^{n} X_i$. If X is intersective for S, then there exists $T \in \mathcal{DS}(S)$ and $i \in \{1, 2, \ldots, n\}$ such that X_i is intersective for T.

(ii) If F is a finite subset of $\mathbb{Z}^+$ and X is intersective for S, then $X \setminus F$ is intersective for S.

(iii) *If F is a finite subset of $\mathbb{Z}^+$ and X is intersective for S, then X is intersective for $S \setminus F$.*

Proof. We first prove (i). Since we have

$$X = X_1 \sqcup \left(\bigsqcup_{i=2}^{n} X_i \right),$$

by induction on n it suffices to prove the result for $n = 2$. Hence, we will let $X = Y \sqcup Z$. Assume, for a contradiction, that neither Y nor Z is intersective for any $T \in \mathcal{DS}(S)$.

Since $S \in \mathcal{DS}(S)$ we are assuming, in particular, that Y is not intersective for S. Hence, there exists $T_1 \in \mathcal{DS}(S)$ such that $FS(T_1) \cap Y = \emptyset$. Since $T_1 \in \mathcal{DS}(S)$, and we are assuming that Z is not intersective for T_1, there exists $T_2 \in \mathcal{DS}(T_1)$ such that $FS(T_2) \cap Z = \emptyset$. By Lemma 2.58 (i), we have $FS(T_2) \subseteq FS(T_1)$. Hence,

$$FS(T_2) \cap (Y \sqcup Z) = \emptyset.$$

By Lemma 2.58 (ii), we have $T_2 \in \mathcal{DS}(S)$. Since X is intersective for S we must have $FS(T_2) \cap X \neq \emptyset$, contradicting our deduction that $FS(T_2) \cap (Y \sqcup Z) = \emptyset$.

Moving on to the proof of (ii), we have $X = F \sqcup (X \setminus F)$. Let $m = \max(f \in F)$, take any $C \in \mathcal{DS}(S)$, and define $D = C \setminus [1, m]$. Clearly, $D \in \mathcal{DS}(S)$. However, $FS(D) \cap F = \emptyset$ since all elements of D are larger than any element of F. Hence, F is not intersective for S. By part (i), this means $X \setminus F$ must be intersective for S.

Part (iii) follows from the observation that if $T \in \mathcal{DS}(S \setminus F)$ then $T \in \mathcal{DS}(S)$ and by definition $FS(T) \cap X \neq \emptyset$. □

The next lemma is the beginning of how we prove the existence of a monochromatic infinite sumset.

Lemma 2.61. *Let X be intersective for S. Then, for any $T \in \mathcal{DS}(S)$ there exists a finite set $E \subseteq T$ such that the following holds:*

for every $x \in FS(T \setminus E)$ there exists $d \in FS(E)$ such that $x + d \in X$.

Proof. Let $T \in \mathcal{DS}(S)$ be arbitrary and assume, for a contradiction, that no such E exists. Pick $x_0 \in T$ arbitrarily and let E_0 consist of those elements from T that sum to x_0 (there may be more than one way to sum to x_0; include them all so that $x_0 \notin FS(T \setminus E_0)$). Since E_0 is finite, there exists

$$x_1 \in FS(T \setminus E_0)$$

such that for all $d \in FS(E_0)$ we have $x_1 + d \notin X$. Let $S(x_1)$ be those elements from $T \setminus E_0$ that sum to x_1 and note that $S(x_1)$ is finite. Let

$$E_1 = E_0 \cup S(x_1).$$

Now, for $i = 2, 3, \ldots$, recursively pick

$$x_i \in FS(T \setminus E_{i-1})$$

and determine $S(x_i) \subseteq T \setminus E_{i-1}$ so that for all $d \in FS(E_{i-1})$ we have $x_i + d \notin X$. Note that by construction $x_i \notin \{x_0, x_1, \ldots, x_{i-1}\}$. To conclude the recursive step, define

$$E_i = E_{i-1} \cup S(x_i).$$

We now have, for any $n \in \mathbb{Z}^+$, elements $x_0, x_1, x_2, \ldots, x_n$ such that $x_n + d \notin X$ for any $d \in FS(E_{n-1})$. In particular, $x_n + d \notin X$ for any $d \in FS(\{x_0, x_1, \ldots, x_{n-1}\})$ since the elements of T used to obtain x_i and x_j are, by construction, distinct for $i \neq j$.

Consider

$$Y = \{x_{2i} + x_{2i-1} : i \in \mathbb{Z}^+\}.$$

By construction, $Y \in \mathcal{DS}(S)$, so $FS(Y) \subseteq FS(S)$. Let $y \in FS(Y)$ be arbitrary. Write y as

$$y = x_{2a} + \left(x_{2a-1} + \sum_{i \in I} (x_{2i} + x_{2i-1}) \right) = x_{2a} + d,$$

for some $I \subseteq [1, a-1]$.

Note that $d \in FS(E_{2a-1})$ so that $x_{2a} + d \notin X$ by construction. Hence, $y \notin X$. Since $y \in FS(Y)$ is arbitrary, we have $FS(Y) \cap X = \emptyset$. Since $Y \in \mathcal{DS}(S)$ this means that X is not intersective for S, a contradiction. □

Using this lemma leads us to the following result.

Lemma 2.62. *Let X be intersective for S. Then, for any $T \in \mathcal{DS}(S)$ there exists $d \in FS(T)$ such that*

$$X_d = \{x \in X : x + d \in X\}$$

is intersective for some $D \in \mathcal{DS}(S)$.

Proof. Let $T \in \mathcal{DS}(S)$ be arbitrary and take $E \subseteq T$ as in Lemma 2.61. Consider

$$Y = X \cap FS(T \setminus E).$$

We will show that Y is intersective for $T \setminus E$. To this end, let

$$C \in \mathcal{DS}(T \setminus E)$$

be arbitrary. By part (i) of Lemma 2.58 we have $FS(C) \subseteq FS(T \setminus E)$. Hence, in order to show that Y is intersective for $T \setminus E$ we need only show that $FS(C) \cap X \neq \emptyset$.

From part (ii) of Lemma 2.58, since $C \in \mathcal{DS}(T \setminus E)$ and $T \setminus E \in \mathcal{DS}(S)$

we have $C \in \mathcal{DS}(S)$. Since X is intersective for S we have $FS(C) \cap X \neq \emptyset$, thereby showing that

$$Y \text{ is intersective for } T \setminus E.$$

Now, by choice of E we have that for every $x \in FS(T \setminus E)$ there exists $d \in FS(E)$ such that $x + d \in X$. Hence,

$$Y = X \cap FS(T \setminus E) = \bigcup_{d \in FS(E)} X_d.$$

Since E is finite, we see that the union is finite.

We can conclude that

$$Y = \bigsqcup_{i=1}^{n} Y_i$$

where

$$Y_i = (X_{d_i} \cap FS(T \setminus E)) \setminus \bigsqcup_{j=1}^{i-1} Y_j$$

and $\{d_1, d_2, \ldots, d_n\} \subseteq FS(E)$. By part (i) of Lemma 2.60, there exist $D \in \mathcal{DS}(T \setminus E)$ and $j \in [1, n]$ such that Y_j is intersective for D. As $X_{d_j} \supseteq Y_j$ we see that for some $d_j \in FS(E)$, the set X_{d_j} is intersective for D. Noting that $d_j \in FS(T)$ and applying part (ii) of Lemma 2.58 finishes the proof. □

The next lemma puts everything together and is the essence of the proof of Hindman's Theorem. After this lemma, a simple observation will complete the proof. In the proof of Lemma 2.63 we will be using the following notation, which we informally introduced in the proof of Lemma 2.61.

Notation. Let $T \in \mathcal{DS}(S)$. For $t \in T$, let $S(t)$ be those elements of S that sum to t (if there is more than one way to sum to t, include all possible ways).

Lemma 2.63. *Let X be intersective for S. Then there exists $A \in \mathcal{DS}(S)$ such that $FS(A) \subseteq X$.*

Proof. We iteratively apply Lemma 2.62. Let $X_0 = X$ and $D_0 = S$ and pick $T_0 \in \mathcal{DS}(D_0)$. By Lemma 2.62 there exists $d_0 \in FS(T_0)$ such that

$$X_1 = \{x \in X_0 : x + d_0 \in X_0\}$$

is intersective for some $C_1 \in \mathcal{DS}(D_0)$.

From C_1 we remove any element c with $S(c) \cap S(d_0) \neq \emptyset$ and note that there are only a finite number of such $c \in C_1$ as only one element in C_1 can use any given element of the finite set $S(d_0)$. We let D_1 be the remaining set:

$$D_1 = C_1 \setminus \{c \in C_1 : S(c) \cap S(d_0) \neq \emptyset\}.$$

By part (iii) of Lemma 2.60 we have that X_1 is intersective for D_1.

We now take $T_1 \in \mathcal{DS}(D_1)$. Again, by Lemma 2.62, there exists $d_1 \in FS(T_1)$ such that

$$X_2 = \{x \in X_1 : x + d_1 \in X_1\}$$

is intersective for some $C_2 \in \mathcal{DS}(D_1)$. Letting

$$D_2 = C_2 \setminus \{c \in C_2 : S(c) \cap (S(d_1) \cup S(d_0)) \neq \emptyset\}$$

we have, by part (iii) of Lemma 2.60, that X_2 is intersective for D_2.

Iterating the above deductions, we take $T_i \in \mathcal{DS}(D_i)$. Then there exists $d_i \in FS(T_i)$ such that

$$X_{i+1} = \{x \in X_i : x + d_i \in X_i\}$$

is intersective for some $C_{i+1} \in \mathcal{DS}(D_i)$. Letting

$$D_{i+1} = C_{i+1} \setminus \{c \in C_{i+1} : S(c) \cap \bigcup_{j=0}^{i} S(d_j) \neq \emptyset\}$$

we see that X_{i+1} is intersective for D_{i+1}.

From this iterative construction, here is a summary of what we have shown so far:

- $T_i \in \mathcal{DS}(D_i)$ so that $T_i \subseteq FS(D_i)$ and $d_i \in FS(D_i)$;
- $X_{i+1} \subseteq X_i$ and $D_{i+1} \subseteq FS(D_i)$;
- $x + d_i \in X_i$ for all $x \in X_i$ and hence for all $x \in X_{i+1}$;
- $S(d_i) \cap \bigcup_{j=0}^{i-1} S(d_j) = \emptyset$.

Let $D = \{d_i\}_{i \in \mathbb{Z}^+}$. As D is an infinite set of distinct positive integers, it contains an infinite increasing subsequence. Thus, we may assume that $d_0 < d_1 < \cdots$. By the last bullet point above we see that $D \in \mathcal{DS}(S)$. Hence, we now define $a_n \in FS(D)$ inductively by starting with $a_0 \in FS(D) \cap X$ arbitrarily (which exists since X is intersective for S). Let k_j (for $j \in \mathbb{Z}^+$) be the maximal index k satisfying

$$d_k \leq \sum_{i=0}^{j-1} a_i$$

and take

$$a_j \in X_{k_j+1} \cap FS(D).$$

We now show that $A = \{a_j\}_{j \in \mathbb{Z}^+}$ satisfies the lemma. First, by construction, $A \in \mathcal{DS}(S)$. We need to show that $FS(A) \subseteq X$. Let

$$a = \sum_{j=0}^{n+1} a_{i_j}$$

represent an arbitrary element of $FS(A)$. For ease of exposition, let $M = i_{n+1}$. Considering

$$a - a_M = \sum_{j=0}^{n} a_{i_j}$$

we have that each a_{i_j} is the sum of elements from D unique to a_{i_j} (see the fourth bullet point above) so that we may write

$$a - a_M = \sum_{j=0}^{m} d_{\ell_j}$$

for some $m \in \mathbb{Z}^+$. By construction, we know that

$$\sum_{j=0}^{n} a_{i_j} < d_{k_M+1}.$$

Hence, $\ell_m < k_{M+1}$.

We have, by bullet point 2, that

$$a_M \in X_{k_M+1} \subseteq X_{k_M} \subseteq \cdots \subseteq X_{\ell_m}.$$

Hence, by bullet point 3, we see that

$$a_M + d_{\ell_m} \in X_{\ell_m} \subseteq X_{\ell_{m-1}+1} \subseteq X_{\ell_{m-1}}.$$

Applying bullet point 3 again, we have

$$(a_M + d_{\ell_m}) + d_{\ell_{m-1}} \in X_{\ell_{m-1}} \subseteq X_{\ell_{m-2}+1} \subseteq X_{\ell_{m-2}}.$$

Continuing this, we find that

$$a = a_M + \sum_{j=0}^{m} d_{\ell_j} \in X_{\ell_0} \subseteq X$$

so that $a \in X$. As a is arbitrary, we have $FS(A) \subseteq X$. □

To finish the proof of Hindman's Theorem, note that clearly $\mathbb{Z}^+$ is intersective for $\mathbb{Z}^+$. For any finite coloring of $\mathbb{Z}^+$, we may write $\mathbb{Z}^+ = \bigsqcup_{i=1}^{n} X_i$, where the X_i are the color classes. By part (i) of Lemma 2.60, we have, for some $j \in [1, n]$, that X_j is intersective for some $Y \in \mathcal{DS}(\mathbb{Z}^+)$. By Lemma 2.63, we have the existence of an infinite set $A \in \mathcal{DS}(Y)$ such that $FS(A) \subseteq X_j$. Hence, $FS(A)$ is a monochromatic infinite sumset.

Having Hindman's Theorem under our belt, we can translate to the multiplicative setting via the simple fact that $2^{a+b} = 2^a \cdot 2^b$. We use the following language.

Definition 2.64 (Finite products, Product-set)**.** Let $S = \{s_i\}_{i=1}^{\infty} \subseteq \mathbb{Z}^+$. The set of *finite products* of S, denoted $FP(S)$, is defined by

$$FP(S) = \left\{ \prod_{i \in I} s_i : \emptyset \neq I \subseteq \mathbb{Z}^+,\ |I| < \infty \right\}.$$

We refer to $FP(S)$ as a *product-set.*

Corollary 2.65 (Multiplicative Hindman's Theorem)**.** *Let $r \in \mathbb{Z}^+$. Every r-coloring of $\mathbb{Z}^+$ admits an infinite set $S \subseteq \mathbb{Z}^+$ such that $FP(S)$ is monochromatic.*

As mentioned previously in this chapter (see Section 2.2.3), we still do not know if every finite coloring of $\mathbb{Z}^+$ admits a monochromatic set of the form $\{a, b, a+b, ab\}$ (we only know it is true for 2 colors and that, if it is true for 3 colors, then we need to color $[1, 11706659]$, at least; see the end of [110]). However, thanks to Hindman [110] we have the following result, which we will not prove.

Theorem 2.66. *Let $r \in \mathbb{Z}^+$. Every r-coloring of $\mathbb{Z}^+$ admits infinite sets $B, C \subseteq \mathbb{Z}^+$ such that $FS(B) \cup FP(C)$ is monochromatic.*

Theorem 2.66 was published in 1979 and proofs that we may or may not take $B = C$ have not materialized.

2.5 Density Results

We end this chapter by presenting some analytic approaches to integer Ramsey theory. We will mainly be considering arithmetic progressions. The standard notation used for density results concerning van der Waerden's Theorem is given next.

Notation. For $k, n \in \mathbb{Z}^+$ denote by $r_k(n)$ the maximal size of a subset of $[1, n]$ with no k-term arithmetic progression.

We start with a result due to Behrend [14] that seems to have been overlooked. This was communicated to the author by Tom Brown. Behrend's result appeared just a year after Erdős and Turán [71] defined $r_3(n)$ and showed that

$$\frac{r_3(n)}{n} < \frac{3}{8} + o(1).$$

Behrend's result is more sweeping.

Theorem 2.67. *For each $k \in \mathbb{Z}^+$ we have that*

$$L_k = \lim_{n\to\infty} \frac{r_k(n)}{n}$$

exists and, more importantly,

$$L = \lim_{k\to\infty} L_k \in \{0,1\}.$$

Proof. The fact that L_k exists is standard and left to the reader as Exercise 2.22. The fact that L exists follows from $0 \le L_1 \le L_2 \le \cdots \le L_k \le \cdots \le 1$, which is a monotonic infinite sequence on a closed interval. Thus, we need only show that L is either 0 or 1.

We will first show that, for every $n \in \mathbb{Z}^+$, we have

$$\frac{r_k(n)}{n} > L_k. \tag{2.3}$$

Assume, for a contradiction, that there exists $m \in \mathbb{Z}^+$ such that Inequality (2.3) is false for m. We may assume that for all $n \in \mathbb{Z}^+$ we have

$$\frac{r_k(m)}{m} \le \frac{r_k(n)}{n}. \tag{2.4}$$

We next note that $r_k(mn) \le nr_k(m)$ since if $S \subseteq [1, mn]$ avoids k-term arithmetic progressions, then it is necessary (but not sufficient) that each of the intervals $[1,m], [m+1, 2m], \ldots, [(n-1)m+1, nm]$ contains at most $r_k(m)$ integers from S. From Inequality (2.4) and this observation, we have

$$\frac{r_k(m)}{m} \le \frac{r_k(mn)}{mn} \le \frac{nr_k(m)}{mn} = \frac{r_k(m)}{m}$$

and we conclude that $r_k(mn) = nr_k(m)$ for all $n \in \mathbb{Z}^+$.

Define the intervals

$$A_i = [(i-1)m + 1, im], \quad i = 1, 2, \ldots, w(k; 2^m),$$

where $w(k; 2^m)$ is the van der Waerden number. Consider

$$n = w(k; 2^m).$$

Letting S be a set of $r_k(mn)$ elements from $[1, mn]$ with no k-term arithmetic progression, we conclude that each of the intervals A_i must contain exactly $r_k(m)$ elements of S.

Define the 2^m-coloring of $[1, n]$ by coloring $i \in [1, n]$ with $(\alpha_1, \alpha_2, \ldots, \alpha_m)$, where $\alpha_j = 1$ if the jth element of interval A_i is in S; let $\alpha_j = 0$ otherwise. Clearly, we have less than 2^m possible colors so that this is a well-defined coloring (each interval contains at least one element, so color $(0, 0, \ldots, 0)$ is not used).

Since $n = w(k; 2^m)$, by van der Waerden's Theorem we have a monochromatic k-term arithmetic progression in $[1, n]$ under this coloring. This means that we have intervals

$$A_a, A_{a+d}, A_{a+2d}, \ldots, A_{a+(k-1)d},$$

where the elements of S occupy the same relative positions in each interval by definition of the coloring. We conclude that S contains a k-term arithmetic progression by taking the first element in each of these intervals, a contradiction. Hence, we conclude that Inequality (2.3) holds.

We now turn to another sequence which also has limit L, namely the upper density (see Definition 1.8). Let $\overline{d}_k$ be chosen so that all subsets of $\mathbb{Z}^+$ with upper density greater that $\overline{d}_k$ contain a k-term arithmetic progression while, for any $\epsilon > 0$, there exists a set with upper density $\overline{d}_k - \epsilon$ that does not contain a k-term arithmetic progression. Then $\overline{d}_k \leq L_k$ since, otherwise, with $\epsilon = \frac{1}{2}(\overline{d}_k - L_k) > 0$ we have $\overline{d}_k - \epsilon > L_k$, a contradiction.

Next, we show that $L_k \leq \overline{d}_{k+1}$. Define the auxiliary sequence $\{x_i\}$ by $x_1 = 4$ and $x_{i+1} = x_i!$ for $i \in \mathbb{Z}^+$. For each x_j, let

$$s_1^{(j)} < s_2^{(j)} < \cdots < s_{r_k(x_j)}^{(j)}$$

be a sequence of $r_k(x_j)$ elements of $[1, x_j]$ with no k-term arithmetic progression. For each $j > 1$, from the sequence $\{s_i^{(j)}\}$ remove any term less than $2(x_{j-1} + 1)$ so that there exists c_j such that

$$2(x_{j-1} + 1) \leq s_{c_j}^{(j)} < s_{c_j+1}^{(j)} < \cdots < s_{r_k(x_j)}^{(j)}. \tag{2.5}$$

Note that the number of terms in each sequence in (2.5) is at least

$$r_k(x_j) - 2x_{j-1} - 1.$$

By definition of x_j, we have

$$\lim_{j \to \infty} \frac{x_{j-1}}{x_j} = 0$$

so that the number of terms in each sequence is

$$r_k(x_j) - o(x_j),$$

meaning that for any $\epsilon > 0$ we have a set of density $\frac{r_k(x_j)}{x_j} - \epsilon$ elements with no arithmetic progression of length k provided j is sufficiently large. We use this to tie $r_k(x_j)$ and $\overline{d}_k$ together.

Consider

$$S = \bigsqcup_{j=1}^{\infty} \left\{ s_{c_j}^{(j)}, s_{c_j+1}^{(j)}, \ldots, s_{r_k(x_j)}^{(j)} \right\},$$

where $c_1 = 1$, and note that

$$s_{c_{j+1}}^{(j+1)} - s_{r_k(x_j)}^{(j)} > 2x_j + 1. \tag{2.6}$$

Because of this, the longest arithmetic progression in S can contain only k terms. To see this, let

$$\left\{s_{c_j}^{(j)}, s_{c_j+1}^{(j)}, \ldots, s_{r_k(x_j)}^{(j)}\right\} \tag{2.7}$$

contain $\ell > 1$ terms of an arithmetic progression. Note that the common difference of this arithmetic progression is between 1 and $\frac{x_j}{\ell-1}$. From Inequality (2.6) we see that our arithmetic progression cannot contain any term larger than $s_{r_k(x_j)}^{(j)}$. For this same reason, the arithmetic progression may only contain at most one element of S that is less than $s_{c_j}^{(j)}$ (otherwise the common difference is constrained by $\frac{x_{j-1}}{\ell-1}$ and cannot contain any term from the elements in (2.7)).

By definition of $r_k(x_j)$ we have $\ell \leq k-1$ and hence we conclude that any arithmetic progression in S has at most k terms. Thus, for any $\epsilon > 0$, we have a set with upper density at least $L_k - \epsilon$ with no $(k+1)$-term arithmetic progression. From this we conclude that $L_k \leq \overline{d}_{k+1}$. So, for all $k \in \mathbb{Z}^+$, we now have

$$\overline{d}_k \leq L_k \leq \overline{d}_{k+1} \tag{2.8}$$

and we conclude that

$$\lim_{k\to\infty} \overline{d}_k = L.$$

Finally, we are in a position to show that L is either 0 or 1. Clearly $0 \leq L \leq 1$. Assume, for a contradiction that $0 < L < 1$. Since $\lim_{k\to\infty} \frac{L_{k-1}}{L_k} = 1$, choose k so that

$$L_{k-1} > L \cdot L_k$$

and let $\epsilon > 0$ be chosen so that

$$4\epsilon < L_{k-1} - L \cdot L_k.$$

Let N be chosen so that for all $n \geq N$, every sequence of more that $(L_k + \epsilon)n$ integers in $[1, n]$ contains a k-term arithmetic progression.

Define the intervals

$$B_i = [(i-1)N + 1, iN], \quad i \in \mathbb{Z}^+.$$

Let $T \subseteq \mathbb{Z}^+$ be a set of upper density at least $L_{k-1} - \frac{\epsilon}{2}$ that contains no arithmetic progression of length k, which is possible by the right inequality in Inequality (2.8).

As we did with the intervals A_i, we assign B_i one of 2^N colors based on the positions of the elements of T in B_i: color B_i with $(\alpha_1, \alpha_2, \ldots, \alpha_N)$ where $\alpha_j = 1$ if the jth element of interval B_i is in T; let $\alpha_j = 0$ otherwise. Unlike when coloring the interval A_i, the color $(0, 0, \ldots, 0)$ may be used for some of the B_i, indicating an interval containing no element of T. We must address this issue.

Let γ be the lower density of the set of indices on B_i for which $T \cap B_i = \emptyset$. Call these intervals *empty intervals*. Choose $M > N$ so that we have more than $(\gamma - \epsilon)M$ empty intervals among $B_1, B_2, \ldots, B_M$ but at least $(L_k - \epsilon)MN$ elements of S in $[1, MN]$.

Since we have at most $(1 - \gamma + \epsilon)M$ intervals that are not empty intervals, and each such interval has at most $(L_k + \epsilon)N$ elements of T (else the interval contains a k-term arithmetic progression), we see that

$$(1 - \gamma + \epsilon)(L_k + \epsilon) \geq L_{k-1} - \epsilon.$$

This gives us

$$\gamma \leq \frac{(1 + \epsilon)(L_k + \epsilon) - L_{k-1} + \epsilon}{L_k + \epsilon} < \frac{L_k - L_{k-1} + 4\epsilon}{L_k}.$$

By choice of ϵ we have

$$\gamma < 1 - L$$

and conclude that the number of indices on B_i for which B_i is not an empty interval has upper density more than L.

Let $w = w(k; 2^N)$ and note that $L_w \leq L$ so that, among these intervals that are not empty intervals we have a w-term arithmetic progression of intervals:

$$B_a, B_{a+d}, B_{a+2d}, \ldots, B_{a+(w-1)d}.$$

Going back to the color assigned to each interval, we see that we have a 2^N-coloring of an arithmetic progression of length $w(k; 2^N)$. By van der Waerden's Theorem we have a monochromatic k-term arithmetic progression among the indices. Hence, we have

$$B_{a'}, B_{a'+d'}, \ldots, B_{a'+(k-1)d'}$$

with the property that the elements of T occupy the same relative positions in each interval. We conclude that T contains an arithmetic progression of length k. But this contradicts our choice of T, thereby completing the proof. □

2.5.1 Roth's Theorem

Notation. In this subsection, i is reserved for $i = \sqrt{-1}$ and for $c \in \mathbb{C}$ we use $\overline{c}$ to represent the complex conjugate.

We will be doing analysis over the group $(\mathbb{Z}_n, +)$: for $f : \mathbb{Z}_n \to \mathbb{C}$, let the (discrete) Fourier transform of f be denoted by $\widehat{f}$ and given by (for $t \in \mathbb{Z}_n$):

$$\widehat{f}(t) = \frac{1}{n} \sum_{k=0}^{n-1} f(k)\overline{\chi(kt)}.$$

Instead of appealing to van der Waerden's Theorem as was done in the proof of Theorem 2.67, our goal is now to prove arithmetic progressions exist in certain sets. We start with the simplest case to consider: 3-term arithmetic progressions. This does not mean that the result concerning these is easy to prove. In 1952, Roth [174] proved that any subset of $\mathbb{Z}^+$ with positive upper density must contain a 3-term arithmetic progression. Note that since every finite coloring of $\mathbb{Z}^+$ must have at least one color with positive upper density, Roth's Theorem is stronger than the existence of $w(3;r)$ for all $r \in \mathbb{Z}^+$.

Theorem 2.68 (Roth's Theorem). *For any $\epsilon > 0$, there exists $N = N(\epsilon)$ such that for any $n \geq N$, if $A \subseteq [1,n]$ with $|A| > \epsilon n$, then A contains a 3-term arithmetic progression.*

Remark. Another way of stating Roth's Theorem is: $r_3(n) = o(n)$.

Since we are dealing with 3-term arithmetic progressions, we can consider integer solutions to $x+y=2z$, where x, y, and z are not all equal. To effectively use discrete Fourier analysis, instead of looking for solutions in $\mathbb{Z}^+$, we will be looking at solutions in $\mathbb{Z}_n$, so we are considering solutions to $x+y \equiv 2z \pmod{n}$. A simple observation will allow us to translate back to $\mathbb{Z}^+$.

The general approach of the proof is to consider cases depending on the sizes of the Fourier transform coefficients. It is best to have a little intuition into what these mean. Consider the recovery formula given in part (i) of Theorem 1.9:

$$\begin{aligned} f(j) &= \sum_{k=0}^{n-1} \widehat{f}(k)\chi(kj) \\ &= \sum_{k=0}^{n-1} \widehat{f}(k)e^{\frac{2\pi i kj}{n}} \\ &= \sum_{k=0}^{n-1} \widehat{f}(k)\cos\left(\frac{2\pi kj}{n}\right) + i\sum_{k=0}^{n-1} \widehat{f}(k)\sin\left(\frac{2\pi kj}{n}\right) \\ &= \widehat{f}(0) + \sum_{k=1}^{n-1} \widehat{f}(k)\cos\left(\frac{2\pi kj}{n}\right) + i\sum_{k=1}^{n-1} \widehat{f}(k)\sin\left(\frac{2\pi kj}{n}\right). \quad (2.9) \end{aligned}$$

The first observation is that $\widehat{f}(0)$ is a constant term in $f(j)$, while the other values of $\widehat{f}(k)$ are contracted by a factor depending on j. To guide our intuition, we then couple this with the fact that any periodic function over $\mathbb{R}$ (like $\cos(x)$) has roots in arithmetic progression (not necessarily of integers).

First, we need to recall how we translate from a set $A \subseteq [1,n]$ to a function $f : \mathbb{Z}_n \to \mathbb{C}$ by using the indicator function:

$$A(j) = \begin{cases} 1 & \text{if } j \in A; \\ 0 & \text{if } j \notin A. \end{cases} \quad (2.10)$$

Hence, we can move from a set A to its associated function $A(j)$ easily. We also note that

$$\widehat{A}(0) = \frac{|A|}{n},$$

i.e., the density of A in $[1, n]$.

Now, looking at our recovery formula in Equation (2.9), first consider the case when all $\widehat{A}(k), k \neq 0$, are "small." Then the Fourier coefficients are spread out across all periodic parts. So, A can be viewed as behaving like a random set. If you are familiar with sound waves and Fourier series, you could associate A with the sound of white noise (for n very large).

Now consider the situation when one $\widehat{A}(k), k \neq 0$, is "large." This means that we have a higher-than-random concentration along some arithmetic progression determined by trigonometric functions. As for the sound wave analogy, you could associate A with a sound where you are able to pick out a particular tone (for n very large).

Consider the graphs in Figures 2.3 and 2.4 of the real part of Equation (2.9), treating j as a real variable (so that j is represented on the x-axis).

When viewing the figures, the graph intersects the x-axis precisely at those integers not in the set, while the graph attains a value of 1 precisely at those integers which are in the set.

Since cosine is a periodic function, if one particular instance of a cosine has a large coefficient in Equation (2.9) then that wave will have a dominant presence in the graph (of course, this is only true visually in extreme circumstances and/or with small sets).

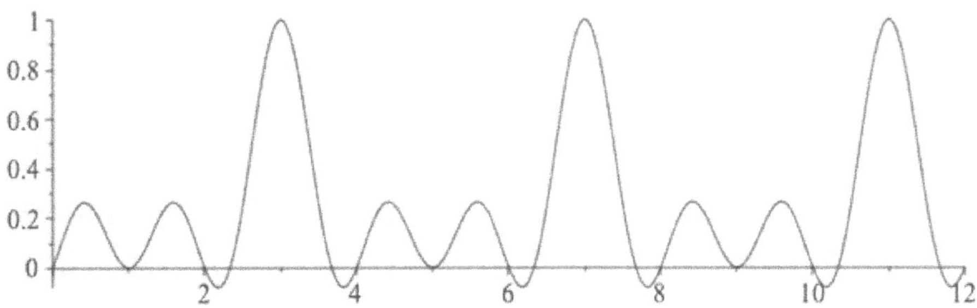

FIGURE 2.3
The real part of the recovery formula for the discrete Fourier transform of a set with all elements in arithmetic progression: the set considered is $\{3, 7, 11\} \subseteq [1, 12]$

In Figure 2.3 we see that we have equally spaced peaks at 3, 7, and 11 (representing the set). Focusing on the wave making these peaks, there would be some period P such that this wave is coming from cosines with period P (and some multiples of P to amplify the wave). The Fourier transform coeffi-

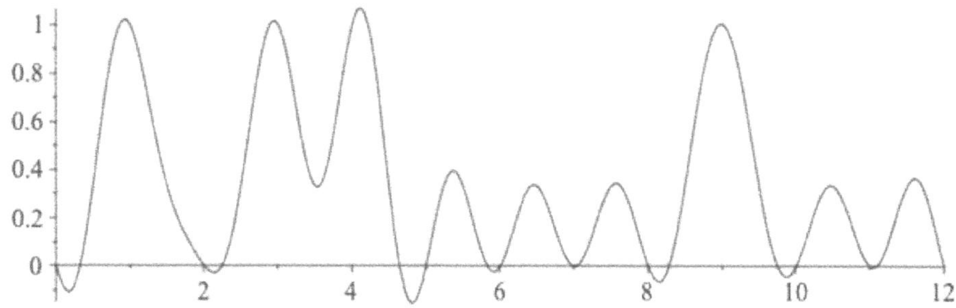

FIGURE 2.4
The real part of the recovery formula for the discrete Fourier transform of a set with randomly chosen elements: the set considered is $\{1,3,4,9\} \subseteq [1,12]$

cients for this particular example are either 0 or .25 (where .25 is considered "large," especially since the density of the set is .25).

In Figure 2.4 we do not have an apparent wave running through the set. The Fourier transform coefficients for this particular example are at most $\frac{1}{6}$ while the density of the set is twice that. So we may consider these coefficients "small" for intuition purposes.

The actual degree of "smallness" and "largeness" is directed as needed in the proof of Roth's Theorem and is based on very large sets. As the graphs in Figures 2.3 and 2.4 are based on subsets of $[1,12]$ they are only provided to help aid intuition as to how the arithmetic progression content of a set is encoded in its Fourier transform.

The approach Roth used to prove Theorem 2.68 can now be expanded upon. If one Fourier transform coefficient $\widehat{A}(k)$ is "large" then there exists an arithmetic progression P_1 such that the density of $A \cap P_1$ in P_1 is larger than the density of A in $[1,n]$. Via the obvious correspondence between P_1 and $[1,|P_1|]$ we now consider $A_1 = A \cap P_1$ as a set in $[1,|P_1|]$ where the density of A_1 in $[1,|P_1|]$ is greater than the density of A in $[1,n]$. If we again have a large Fourier transform coefficient we repeat this argument. Provided that the increase in density is greater than some constant (allowing dependency on the original density), we either discover a desired arithmetic progression or we eventually get to a density greater than 1, which is a contradiction. Hence, we can assume that all Fourier transform coefficients are "small."

If all coefficients are small, A can be viewed as similar to a random set, i.e., we can consider $\mathbb{P}(x \in A) \approx \delta$ for all $x \in [1,n]$, where δ is the density of A in $[1,n]$. Let's investigate what happens in the random setting. Order all 3-term arithmetic progressions in $[1,n]$ and let X_i be the indicator random variable with $X_i = 1$ only when the ith arithmetic progression lies in A. We know there are approximately $\frac{n^2}{4}$ arithmetic progressions with 3 terms (hopefully you already did Exercise 1.8) in $[1,n]$. Using the linearity of expectation (see

Lemma 6.1) we have

$$\mathbb{E}\left(\sum_{i=1}^{\frac{n^2}{4}} X_i\right) = \sum_{i=1}^{\frac{n^2}{4}} \mathbb{E}(X_i) = \sum_{i=1}^{\frac{n^2}{4}} \mathbb{P}(X_i = 1) \approx \frac{n^2\delta^3}{4}.$$

Accounting for the n arithmetic progressions with common difference 0 (i.e., the trivial arithmetic progressions), if we have $\frac{n^2\delta^3}{4} - n > 1$, for which

$$n > \frac{8}{\delta^3}$$

suffices, then we expect at least one non-trivial 3-term arithmetic progression to be in A. In other words, as long as $\delta > 0$, for n sufficiently large, a random set of density δ in $[1, n]$ is expected to contain a 3-term arithmetic progression. Of course we have to move from "expected" to "must" in the actual proof.

The last piece of the proof concerns how we go from working in $\mathbb{Z}_n$ (needed so that we may use discrete Fourier analysis tools) to justifying the result on $\mathbb{Z}^+$. If we can show that the solution to

$$x + y \equiv 2z \pmod{n}$$

occurs with $x, z \in \left[\frac{n}{3}, \frac{2n}{3}\right]$ then we know that $x + y = 2z$ is satisfied, i.e., that we have a 3-term arithmetic progression in $\mathbb{Z}^+$.

We are now ready to present a formal proof of Roth's Theorem. We partly follow the exposition given in [136] and will use the notation in Equation (2.10) as well as

$$\chi(t) = e^{\frac{2\pi i t}{n}}$$

($t \in \mathbb{Z}_n$) to represent additive characters of $\mathbb{Z}_n$.

Proof of Theorem 2.68. Let $A \subseteq [1, n]$ with $|A| = \delta n$. Define

$$B = A \cap \left[\frac{n}{3}, \frac{2n}{3}\right].$$

We may assume that

$$|B| \geq \frac{|A|}{5}; \tag{2.11}$$

otherwise, one of $[1, \frac{n}{3}]$ and $[\frac{2n}{3}, n]$ contains at least $\frac{2}{5}|A|$ elements. This gives us an interval of length $\frac{n}{3}$ that contains at least $\frac{2}{5}\delta n$ elements of A meaning that the relative density of a subset of A in an interval I of length $\frac{n}{3}$ is $\frac{6}{5}\delta$. We then repeat this process by partitioning I into three equal intervals. If Inequality (2.11) is never satisfied then we can find a subset of A with relative density $\left(\frac{6}{5}\right)^k \delta$ in an interval of length $\frac{n}{3^k}$. By choosing k such that

$$\left(\frac{6}{5}\right)^k > \delta^{-1}$$

and letting n be sufficiently large, we get a contradiction by having a subset of A with relative density greater than 1.

Under Inequality (2.11), we now estimate the number of solutions M to $x + y \equiv 2z \pmod{n}$ as follows:

$$\begin{aligned}
M &= |\{(x,y,z) \in A \times A \times A : x + y \equiv 2z \pmod{n}\}| \\
&\geq |\{(x,y,z) \in B \times A \times B : x + y \equiv 2z \pmod{n}\}| && (2.12)\\
&= \sum_{\substack{x,y,z \in \mathbb{Z}_n \\ x+y \equiv 2z \pmod{n}}} B(x)A(y)B(z) && (2.13)\\
&= \frac{1}{n} \sum_{x,y,z \in \mathbb{Z}_n} \sum_{j \in \mathbb{Z}_n} B(x)A(y)B(z)\chi(-j(x+y-2z)) && (2.14)\\
&= \frac{1}{n} \sum_{j \in \mathbb{Z}_n} \left(\sum_{x \in \mathbb{Z}_n} B(x)\chi(-jx) \sum_{y \in \mathbb{Z}_n} A(y)\chi(-jy) \sum_{z \in \mathbb{Z}_n} B(z)\chi(2jx) \right) \\
&= n^2 \sum_{j \in \mathbb{Z}_n} \widehat{B}(j)\widehat{A}(j)\widehat{B}(-2j) && (2.15)\\
&= \frac{|A||B|^2}{n} + n^2 \sum_{\substack{j \in \mathbb{Z}_n \\ j \neq 0}} \widehat{B}(j)\widehat{A}(j)\widehat{B}(-2j). && (2.16)
\end{aligned}$$

Notice that the set in Inequality (2.12) enumerates solutions to $x + y = 2z$ since $x, z \in \left[\frac{n}{3}, \frac{2n}{3}\right]$.

Equation (2.14) translates the sum over $x + y \equiv 2z \pmod{n}$ to a sum without this restriction by using the additive characters. For fixed values of x, y, and z, if $x + y - 2z = c \neq 0$ then the inner sum of Equation (2.14) is

$$\sum_{j \in \mathbb{Z}_n} k\chi(-jc) = k \sum_{j \in \mathbb{Z}_n} \chi(-jc)$$

where k is either 0 or 1. Noting that $e^{jc} = (e^c)^j$ and that c is an integer between 1 and $n-1$ we have

$$k \sum_{j \in \mathbb{Z}_n} \chi(-jc) = k \sum_{j=0}^{n-1} \chi(-c)^j = k \cdot \frac{1 - \chi(-c)^n}{1 - \chi(-c)} = 0.$$

Hence, the inner sum of (2.14) is positive only when $x + y \equiv 2z \pmod{n}$, and when this occurs, the result is as given in Equation (2.13).

Equation (2.15) is simply the definition of the discrete Fourier transform.

Isolating the $j = 0$ term in Equation (2.15) gives us the last equation. We view the sum in Equation (2.16) as the error term with a goal of showing that it is smaller than $\frac{|A||B|^2}{n}$. Hence, we let

$$E = n^2 \sum_{\substack{j \in \mathbb{Z}_n \\ j \neq 0}} \widehat{B}(j)\widehat{A}(j)\widehat{B}(-2j)$$

and consider the size of the Fourier transform coefficients. Since we are attempting to show that E is small, we first consider the situation when all Fourier transform coefficients are small (meaning that our set behaves like a random set) as that should be easier.

`Claim 1.` *If $|\widehat{A}(j)| \leq \epsilon$ with $\epsilon \leq \frac{\delta^2}{10}$ for all $j \in \mathbb{Z}_n \setminus \{0\}$, then A contains a 3-term arithmetic progression.*

This follows from a sequence of inequalities. We have

$$\begin{aligned}
|E| &\leq n^2 \left(\max_{\substack{j \in \mathbb{Z}_n \\ j \neq 0}} |\widehat{A}(j)| \right) \left| \sum_{\substack{j \in \mathbb{Z}_n \\ j \neq 0}} \widehat{B}(j)\widehat{B}(-2j) \right| \\
&\leq n^2 \left(\max_{\substack{j \in \mathbb{Z}_n \\ j \neq 0}} |\widehat{A}(j)| \right) \left(\sum_{j \in \mathbb{Z}_n} |\widehat{B}(j)|^2 \right)^{\frac{1}{2}} \left(\sum_{j \in \mathbb{Z}_n} |\widehat{B}(-2j)|^2 \right)^{\frac{1}{2}} \qquad (2.17) \\
&= n \left(\max_{\substack{j \in \mathbb{Z}_n \\ j \neq 0}} |\widehat{A}(j)| \right) \left(\sum_{x \in \mathbb{Z}_n} |B(x)|^2 \right)^{\frac{1}{2}} \left(\sum_{x \in \mathbb{Z}_n} |B(-2x)|^2 \right)^{\frac{1}{2}} \qquad (2.18) \\
&\leq \frac{\delta^2}{10} |B| n \\
&\leq \frac{1}{2n} |A||B|^2. \qquad (2.19)
\end{aligned}$$

Inequality (2.17) is the Cauchy-Schwartz Inequality along with the addition of the $j = 0$ term in each sum, while Equality (2.18) uses Plancherel's equation (see part (iii) of Theorem 1.9).

Using Inequalities (2.16) and (2.19), we have

$$M \geq \frac{1}{2n} |A||B|^2.$$

Appealing to Inequality (2.11), and subtracting the δn trivial solutions (when $x = y = z$ with $x, y, z \in A$), we have, for n sufficiently large, that

$$M \geq \frac{\delta^3}{50} n^2 - \delta n > 0.$$

As noted at the beginning, a solution to $x + y \equiv 2z \pmod{n}$ with $x, z \in B$ is a 3-term arithmetic progressions in $\mathbb{Z}^+$. Noting that this is precisely the set in Inequality (2.12) finishes the proof of Claim 1. $\diamond$

By Claim 1, we can now assume that

$$|\widehat{A}(k)| > \frac{\delta^2}{10}$$

for some $k \in \mathbb{Z}_n \setminus \{0\}$.

The estimate provided by Equation (2.16) is not strong enough to deal with the case when we have a "large" Fourier transform coefficient. Recall that the intuition in this situation is that A has a higher-than-overall relative density along some arithmetic progression. However, we can also see that the coefficients we are dealing with are not really that large. As such, what follows is the more intricate part of the proof.

`Claim 2.` *Let $\epsilon > 0$ and let $|A| = \delta n$. If $|\widehat{A}(k)| > \epsilon$ for some $k \in \mathbb{Z}_n \setminus \{0\}$, then there exists an arithmetic progression P in $\mathbb{Z}^+$ of length at least $\frac{\epsilon}{64}\sqrt{n}$ such that*

$$\frac{|A \cap P|}{|P|} > \delta + \frac{\epsilon}{8}.$$

Fix k so that $|\widehat{A}(k)| > \epsilon$.

The proof of Claim 2 follows two steps:

> `Step A.` Prove the existence of a long arithmetic progression P_n in $\mathbb{Z}_n$ for which the density of $A \cap P_n$ in $\mathbb{Z}_n$ is large.
>
> `Step B.` Show that P_n contains a long arithmetic progression P in $\mathbb{Z}^+$ for which the density of $A \cap P$ in $\mathbb{Z}^+$ is still large.

To start Step A we find an arithmetic progression Q_n in $\mathbb{Z}_n$ of length at least $\frac{\sqrt{n}}{8}$ for which

$$\widehat{Q_n}(k) > \frac{|Q_n|}{2n}.$$

To this end, consider a tiling of the grid $[0, n-1] \times [0, n-1]$ by squares of side length $\sqrt{n-1}$ (some squares may be partial). Place all elements of the set

$$S = \{(j, jk) : j \in \mathbb{Z}_n\}$$

on the grid. By the pigeonhole principle (there are $n-1$ squares containing n elements), some square contains at least two elements of S, say (s, sk) and (t, tk) with $s < t$. Then both

$$t - s \leq \sqrt{n-1}$$

and

$$k(t-s) \leq \sqrt{n-1} \tag{2.20}$$

must hold. Let $d = t - s$ and record for later use that

$$d < \sqrt{n}. \tag{2.21}$$

Let Q_n be the $(2\lfloor \frac{\sqrt{n}}{16} \rfloor + 3)$-term arithmetic progression $\pm \ell d, \ell = 0, 1, \ldots, \lfloor \frac{\sqrt{n}}{16} \rfloor + 1$. As

$$\left(\frac{\sqrt{n}}{16} + 1\right)\sqrt{n} < \frac{n}{2}$$

for n sufficiently large, we see that all elements of Q_n are distinct values in $\mathbb{Z}_n$. For ease of reading, we will use

$$|Q_n| = \frac{\sqrt{n}}{8}$$

even though Q_n may be a few terms longer (and $\frac{\sqrt{n}}{8}$ may not be an integer). Hence, we will denote the density of Q_n in $\mathbb{Z}_n$ by

$$\delta_Q = \frac{1}{8\sqrt{n}}$$

To finish Step A, we will show that the desired P_n is a translate of Q_n. In order to show that $A \cap P_n$ is large, we will need to use

$$|\widehat{Q_n}(k)| > \frac{1}{16\sqrt{n}} = \frac{|Q_n|}{2n}. \tag{2.22}$$

To justify this inequality, we have the following sequence of bounds:

$$\begin{aligned}
|\widehat{Q_n}(k) - \delta_Q| &= \left|\frac{1}{n}\sum_{j\in\mathbb{Z}_n}(Q_n(j)\chi(-kj) - Q_n(j))\right| \\
&\leq \frac{1}{n}\sum_{j\in\mathbb{Z}_n}|Q_n(j)| \cdot |(\chi(-kj) - 1)| \\
&\leq \frac{1}{n}\sum_{j\in Q_n}|(\chi(-kj) - 1)| \\
&\leq \frac{|Q_n|}{n}\max_{j\in Q_n}|\chi(-kj) - 1| \\
&= \delta_Q \max_{j\in Q_n}|\chi(-kj) - 1| \\
&< \frac{1}{2}\delta_Q.
\end{aligned}$$

The last step warrants explanation. Using the bound on $j \in Q_n$ and Inequality (2.20) we find that

$$|kj| \le kd\left(\left\lfloor \frac{\sqrt{n}}{16} \right\rfloor + 1\right) \le \sqrt{n-1}\left(\left\lfloor \frac{\sqrt{n}}{16} \right\rfloor + 1\right) \le \frac{n}{16} + \sqrt{n}.$$

For n sufficiently large we have $\frac{n}{16} + \sqrt{n} < \frac{n}{15}$ so that, for any $j \in Q_n$, we have that $\chi(kj)$ is an nth root of unity that is less than $\frac{2\pi}{15}$ radian from the x-axis. Hence,

$$|\chi(-kj) - 1| < \frac{2\pi}{15} < \frac{1}{2}.$$

Now, since $|\widehat{Q_n}(k) - \delta_Q| < \frac{1}{2}\delta_Q$, Inequality (2.22) follows.

We continue by showing the existence of the desired arithmetic progression P_n that has a "large" intersection with A.

A fundamental question at this stage is: How do we use Fourier analysis to prove large intersections? The answer lies in the incredibly deep observation that $1 \cdot 1 = 1$ and $1 \cdot 0 = 0$. So, if S and T are two sets with associated functions $S(j)$ and $T(j)$, then $S(j)T(j) = 1$ if j is an element of both S and T; otherwise $S(j)T(j) = 0$.

As noted above, we will show that we may take P_n as a translate of Q_n. Hence, we want to show that for some $a \in \mathbb{Z}_n$, the sum

$$f(a) = \sum_{j \in \mathbb{Z}_n} A(j) Q_n(a - j)$$

is large. Notice that this is a convolution and so we can appeal to part (iv) of Theorem 1.9, which states that

$$\widehat{f}(a) = n\widehat{A}(a)\widehat{Q_n}(a). \tag{2.23}$$

Our approach is to show that the average value of $\sum_{\ell \in \mathbb{Z}_n} f(\ell)$ is large to conclude that one particular value must be large, too. We can determine this average easily:

$$\begin{aligned}
\frac{1}{n}\sum_{\ell \in \mathbb{Z}_n} f(\ell) &= \frac{1}{n}\sum_{\ell \in \mathbb{Z}_n}\sum_{j \in \mathbb{Z}_n} A(j)Q_n(\ell - j) \\
&= \frac{1}{n}\sum_{j \in \mathbb{Z}_n}\sum_{\ell \in \mathbb{Z}_n} A(j)Q_n(\ell - j) \\
&= \frac{1}{n}\sum_{j \in \mathbb{Z}_n} A(j)\sum_{\ell \in \mathbb{Z}_n} Q_n(\ell - j) \\
&= \frac{|A|}{n}\,|Q_n| \\
&= \delta\,|Q_n|.
\end{aligned}$$

From this we can conclude that some instance of $f(j)$ has at least $\delta|Q_n|$ terms. Unfortunately, this is not strong enough for us to prove Step A, so we must do more work.

We employ a useful trick next by introducing a *balanced function*, which is an auxiliary function with mean 0. For this, we let

$$C(j) = A(j) - \delta$$

and consider the associated convolution

$$g(a) = \sum_{j \in \mathbb{Z}_n} C(j) Q_n(a-j).$$

We then find that

$$\sum_{\ell \in \mathbb{Z}_n} g(\ell) = 0 \tag{2.24}$$

via exactly the same argument as used on $\sum_{\ell \in \mathbb{Z}_n} f(\ell)$ above. This fact will be exploited and is why using the balanced function is useful. The impetus for using the balanced function is to show that some instance of $g(j)$ is "significantly" more than 0 so that $f(j)$ is "large" (compared with $\delta|Q_n|$, a bound that we know is not strong enough).

Since we are using Fourier transforms, we will need to know how the transform of $C(j)$ relates to the transform of $A(j)$. We have:

$$\widehat{C}(j) = \widehat{A}(j) \tag{2.25}$$

for any $j \in \mathbb{Z}_n \setminus \{0\}$. This follows from the fact that $\sum_{k=0}^{n-1} \chi(kx) = 0$ when $x \neq 0$.

From the pointwise estimate given in Theorem 1.9, coupled with the convolution formula (Equation (2.23)), Inequality (2.22), Equation (2.25), and the bound $|\widehat{A}(k)| > \epsilon$, we see that

$$\frac{1}{n} \sum_{j \in \mathbb{Z}_n} |g(j)| \geq |\widehat{g}(k)| = n \left| \widehat{C}(k) \widehat{Q_n}(k) \right| = n|\widehat{A}(k)| \cdot |\widehat{Q_n}(k)| > \epsilon \frac{\sqrt{n}}{16} = \frac{|Q_n|}{2} \epsilon.$$

From this bound and Equation (2.24) we have

$$\frac{1}{n} \sum_{\ell \in \mathbb{Z}_n} (|g(j)| + g(j)) > \frac{|Q_n|}{2} \epsilon.$$

Having a lower bound on the average value allows us to conclude that for some $a \in \mathbb{Z}_n$ we have

$$|g(a)| + g(a) > \frac{|Q_n|}{2} \epsilon$$

so that

$$g(a) > \frac{|Q_n|}{4} \epsilon.$$

Let P_n be the arithmetic progression obtained by translating Q_n by this a and note that $|P_n| = |Q_n|$ in $\mathbb{Z}_n$.

Since

$$g(a) = \sum_{j\in\mathbb{Z}_n} C(j)Q_n(a-j) = \sum_{j\in\mathbb{Z}_n} A(j)Q_n(a-j) - \delta \sum_{j\in\mathbb{Z}_n} Q_n(a-j)$$
$$= |A \cap P_n| - \delta|P_n|$$

we have

$$|A \cap P_n| - \delta|P_n| > \frac{|P_n|}{4}\epsilon$$

so that

$$\frac{|A \cap P_n|}{|P_n|} > \left(\delta + \frac{\epsilon}{4}\right),$$

which is the density bound we desire to finish Step A.

We now move on to Step B, where we lift permutation P_n from $\mathbb{Z}_n$ to a permutation P in $\mathbb{Z}^+$. A key observation is that the common gap in P_n is less than $\sqrt{n}$ (see Inequality (2.21)) and that the length of P_n is also less than $\sqrt{n}$. Thus, the distance between the first term of P_n and the last term of P_n is less that $(\sqrt{n})^2 = n$ meaning that P_n does not wrap around on itself. Hence, viewing P_n in $\mathbb{Z}^+$ we have $P_n = P_1 \cup P_2$ where both are arithmetic progressions in $\mathbb{Z}^+$. To help visualize this, consider the diagram in Figure 2.5, where $P_n = x_1, x_2, x_3, \ldots, x_t$:

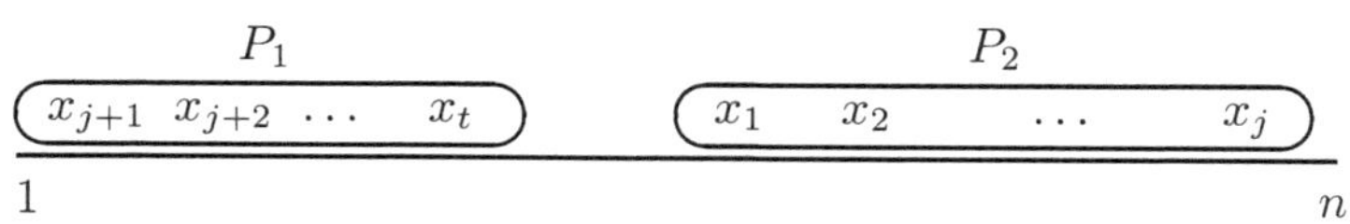

FIGURE 2.5
Lifting P_n from $\mathbb{Z}_n$ to $\mathbb{Z}^+$

We may assume that $|P_1| \leq |P_2|$.

`Case 1.` $|P_1| < \frac{\epsilon}{8}|P_n|$. We have $|A \cap P_2| \geq |A \cap P_n| - |P_1|$ so that in this case $|A \cap P_2| > \left(\delta + \frac{\epsilon}{8}\right)|P_n| \geq \left(\delta + \frac{\epsilon}{8}\right)|P_2|$. Hence,

$$\frac{|A \cap P_2|}{|P_2|} \geq \left(\delta + \frac{\epsilon}{8}\right)$$

and P_2 has length at least

$$\left(1 - \frac{\epsilon}{8}\right)|P_n| = \left(1 - \frac{\epsilon}{8}\right)\frac{\sqrt{n}}{8} > \frac{\epsilon}{64}\sqrt{n}.$$

`Case 2.` $|P_1| \geq \frac{\epsilon}{8}|P_n|$. In this case, each of P_1 and P_2 has at least $\frac{\epsilon}{8}|P_n| =$

$\frac{\epsilon}{64}\sqrt{n}$ elements and A must intersect one of them in at least $\left(\delta + \frac{\epsilon}{4}\right)|P_n|$ places (otherwise the overall density of $A \cap P_n$ in P_n would be strictly less than $\left(\delta + \frac{\epsilon}{4}\right)$, a contradiction). Let $P_3 \in \{P_1, P_2\}$ be that progression with the larger intersection with A. Then,

$$\frac{|A \cap P_3|}{|P_3|} \geq \frac{|A \cap P_3|}{|P_n|} \geq \left(\delta + \frac{\epsilon}{4}\right).$$

We see that in each case, the conditions of Claim 2 are satisfied so that the proof of Claim 2 is complete. ◇

To finish the proof of Roth's Theorem, we start by applying Claims 1 and 2 with $\epsilon = \frac{\delta^2}{10}$. If Claim 1 applies, we are done, so we assume Claim 2 applies. Assuming that A does not contain a 3-term arithmetic progression then neither does $A \cap P$, which has relatively density at least $\delta + \frac{\delta^2}{80}$. Identifying P with $[1, |P|]$ and letting

$$\delta_1 = \delta + \frac{\delta^2}{80}$$

we again apply Claims 1 and 2 with $\delta = \delta_1$ and $\epsilon = \frac{\delta_1^2}{10}$. As before, if Claim 1 applies we are done, so we assume Claim 2 applies again and let the associated arithmetic progression be P_1.

For $k = 2, 3, \ldots$, let

$$\epsilon = \epsilon_k = \frac{\delta_{k-1}^2}{10}$$

and

$$\delta = \delta_k = \delta_{k-1} + \frac{\delta_{k-1}^2}{80}$$

with associated arithmetic progression P_k. Repeatedly applying Claims 1 and 2 with these values of ϵ and δ is valid since P_k can be arbitrarily long. If, for any value of k, Claim 1 applies we are done. Hence, we must determine what happens if Claim 1 never applies.

We have

$$\delta_k = \delta_{k-1} + \frac{\delta_{k-1}^2}{80} = \delta_{k-2} + \frac{\delta_{k-2}^2}{80} + \frac{\delta_{k-1}^2}{80} = \delta_0 + \sum_{i=0}^{k-1} \frac{\delta_i^2}{80},$$

(where δ_0 is the original density of A in $[1, n]$). We know that $\delta_i > \delta_0$ for all $i \in \mathbb{Z}^+$ so we have

$$\delta_k > \delta_0 + k\frac{\delta_0^2}{80}.$$

Hence, choosing

$$k > \frac{80}{\delta_0^2}$$

implies that $\delta_k > 1$, which is not possible. Thus, either Claim 1 must eventually apply or our assumption that A does not contain a 3-term arithmetic progression is incorrect. In either situation, the proof of Roth's Theorem is complete. □

Remark. The curious reader may wonder if a result similar to Roth's Theorem holds for other equations, in particular, for Schur's Theorem. A priori it seems to be a reasonable question: does every set of positive upper density contain a solution to $x+y=z$? Well, if we consider the set of odd integers, we clearly have no solution and have a density of $\frac{1}{2}$. Hence, the question we asked is not the correct one. See [18, 97, 98, 185] for work in this area.

Now that Roth's Theorem gives us $r_3(n) = o(n)$ a question arises: how large can a subset of $[1,n]$ be and still avoid 3-term arithmetic progressions? Roth's Theorem tells us it must have zero upper density; more specifically, Roth's proof informs us that

$$r_3(n) \ll \frac{n}{\log\log n}.$$

We next present a construction due to Behrend [15] (which builds upon work of Salem and Spencer [182]) even though it does not give the best-known lower bound. The current world record for the lower bound on $r_3(n)$ is due to Elkin [61] (Green and Wolf [91] provide a much shorter proof of Elkin's result):

$$r_3(n) \gg \frac{n\log^{\frac{1}{4}} n}{2^{2\sqrt{2\log n}}}. \tag{2.26}$$

The current published world record for the upper bound on $r_3(n)$ is due to Bloom [24] and is

$$r_3(n) \ll \frac{n(\log\log n)^4}{\log n}; \tag{2.27}$$

however, a preprint by Bloom and Sisack [25] gives

$$r_3(n) \ll \frac{n}{(\log n)^{1+c}} \tag{2.28}$$

for some constant $c > 0$.

The idea behind Behrend's proof is to choose a large subset of points on the surface of a high-dimensional sphere with a radius r such that $r^2 \in \mathbb{Z}^+$. Each point will correspond to an integer in a natural way and have the property that if $x+y=2z$, i.e., $z = \frac{x+y}{2}$, then the surface of the sphere must contain two points and their midpoint, a property we know does not hold.

Theorem 2.69. *For n sufficiently large,*

$$r_3(n) > \frac{n}{\sqrt{\log n}\cdot 2^{2\sqrt{2\log n}}}.$$

Proof. Let $k, r^2 \in \mathbb{Z}^+$. For $d \in \mathbb{Z}^+$ and $x \in [1, (2k-1)^d - 1]$ we can write

$$x = \sum_{i=1}^{d} a_i (2k-1)^{i-1}$$

uniquely. We define the map

$$\varphi(x) = (a_1, a_2, \ldots, a_d).$$

Consider those integers j for which $\varphi(j)$ is a d-tuple $(a_1, a_2, \ldots, a_d)$ with both of the following properties:

(i) $0 \le a_i < k$ for $1 \le i \le d$;

(ii) $\displaystyle\sum_{i=1}^{d} a_i^2 = r^2$.

Call this set $S^d(r)$, which is a subset of the integer lattice points on the surface of the $(d-1)$-sphere centered at the origin with radius r.

First, if $x, y, z \in [1, (2k-1)^d - 1]$ satisfy $x + y = 2z$ with $x \neq y$ we will show that we cannot have $\varphi(x), \varphi(y)$, and $\varphi(z)$ all in $S^d(r)$ (i.e., we will provide a formal proof that a sphere cannot contain two points and their midpoint). Assume, for a contradiction, that we do. Using the notation $\|(a_1, a_2, \ldots, a_k)\| = \left(\sum_{i=1}^{k} a_i^2\right)^{\frac{1}{2}}$ (i.e., the standard Euclidean norm), we have

$$2r = \|\varphi(2z)\| = \|\varphi(x+y)\| \le \|\varphi(x)\| + \|\varphi(y)\| = 2r.$$

Since $\|\cdot\|$ is a norm, $\|\varphi(x)\| = \|\varphi(y)\|$, and we have equality in the triangle inequality, we must have $\varphi(x) = \varphi(y)$ meaning that $x = y$, which is not allowed (such a solution corresponds to a trivial 3-term arithmetic progression).

Having shown that $S^d(r)$ is generated by a set of integers with no 3-term arithmetic progression, we determine a large lower bound for the number of points in $S^d(r)$ for some r and d.

There are k^d elements in $A = \{(a_1, a_2, \ldots, a_d) : 0 \le a_i < k\}$. In order for $\mathbf{a} \in A$ to be in $S^d(r)$ we require $\|\mathbf{a}\| = r$. By construction we know that

$$1 \le r^2 \le \sum_{i=1}^{d} (k-1)^2 = d(k-1)^2$$

gives the possible values for r. Since the average value of $|S^d(r)|$ over all possible r values for which $r^2 \in \mathbb{Z}^+$ is

$$\frac{k^d}{d(k-1)^2} > \frac{k^{d-2}}{d},$$

there must exist some value t with $t^2 \in \mathbb{Z}^+$ such that

$$|S^d(t)| > \frac{k^{d-2}}{d}.$$

At this point we see that $\varphi^{-1}(S^d(t))$ is a subset of $[1, (2k-1)^{d-1}]$ with at least $\frac{k^{d-2}}{d}$ elements, no three of which are in arithmetic progression.

Let $n \in \mathbb{Z}^+$ be a sufficiently large given integer. We now take

$$d = \lfloor \sqrt{2 \log n} \rfloor$$

and determine k so that $(2k-1)^d \leq n < (2k+1)^d$. To avoid (what will be) unnecessary negligible terms stemming from the greatest integer (floor) function we will use $d = \sqrt{2 \log n}$ in what follows. Hence, we have

$$2^{d^2} = n^2.$$

So far we have

$$r_3(n) \geq r_3((2k-1)^d - 1) > \frac{k^{d-2}}{d}.$$

Having chosen k so that $n < (2k+1)^d$, this becomes

$$r_3(n) > \frac{(n^{1/d} - 1)^{d-2}}{d\, 2^{d-2}}.$$

Using $2^{d^2} = n^2$, we obtain

$$r_3(n) > \frac{1}{d} \left(\frac{2^{\frac{d}{2}} - 1}{2} \right)^{d-2}.$$

Hence,

$$\begin{aligned} r_3(n) > \frac{1}{d} \left(\frac{2^{\frac{d}{2}} - 1}{2} \right)^{d-2} > \frac{1}{d} \left(2^{\frac{d}{2} - 1 - \frac{1}{d-2}} \right)^{d-2} &= \frac{1}{d} \left(\frac{2^{\frac{d^2}{2}+1}}{2^{2d}} \right) \\ &= \frac{2n}{d 2^{2d}} \\ &= \frac{2n}{\sqrt{2 \log n} \cdot 2^{2\sqrt{2 \log n}}} \\ &> \frac{n}{\sqrt{\log n} \cdot 2^{2\sqrt{2 \log n}}}, \end{aligned}$$

which was to be shown. □

2.5.2 Szemerédi's Theorem

Translating the proof of Roth's Theorem to address longer arithmetic progressions presents the challenge of moving from solving the equation $x + y = 2z$ to solving a system of equations. Clearly, more sophisticated Fourier analysis tools are needed and this was done twenty years later by Roth [175] for 4-term arithmetic progressions. However, this was after Szemerédi [195] proved the result via combinatorial methods. Szemerédi then went on to prove his famous theorem for arbitrarily long arithmetic progressions in [196]:

Theorem 2.70 (Szemerédi's Theorem). *For any $k \in \mathbb{Z}^+$ and any subset $A \subseteq \mathbb{Z}^+$, if $\limsup_{n\to\infty} \frac{|A\cap[1,n]|}{n} > 0$, then A contains a k-term arithmetic progression. Equivalently, given $\epsilon > 0$ there exists an integer $N(k,\epsilon)$ such that for all $n > N(k,\epsilon)$, if $|A| > \epsilon n$, then A contains a k-term arithmetic progression.*

Szemerédi's proof has been referred to as elementary. This does not mean easy. It is better referred to as "from first principles," as Szemerédi felt just the logic of the proof warranted a flow chart. This chart is a directed graph on 24 vertices with 36 edges.

The reader is encouraged to work through this combinatorial feat. In this book, we prove Szemerédi's Theorem in Section 5.2.1, relying on results that we will not prove.

It is worth noting here that Szemerédi's Theorem is equivalent, by Theorem 2.67, to the following theorem, the proof of the equivalence being left to the reader as Exercise 2.23. Note also that Theorem 2.70 settles the value of L in Theorem 2.67.

Theorem 2.71. *For all $k \in \mathbb{Z}^+$, there exists a minimal integer $N(k)$ such that for every partition of $\{1, 2, \ldots, n\} = A \sqcup B$ with $|A| \geq |B|$ and $n \geq N(k)$, the set A contains a k-term arithmetic progression.*

It is interesting to notice that van der Waerden's Theorem can be very similarly stated as: For all $k \in \mathbb{Z}^+$, there exists a minimal integer $M(k)$ such that for every partition of $\{1, 2, \ldots, m\} = A \sqcup B$ with $m \geq M(k)$, one of A and B contains a k-term arithmetic progression.

In this section, we will investigate a closely related result that, on the surface, seems as if it may be sufficient to prove Szemerédi's Theorem. One possible thought about sets satisfying the condition in Theorem 2.70 is that they cannot have arbitrarily long gaps (forewarning: this is not true) or, if they do, then there must be arbitrarily long intervals where they don't (forewarning: this is not true either). These seem reasonable, so we will investigate what we can say about this situation.

Definition 2.72 (Syndetic, Piecewise syndetic). Let $S = \{s_i\}_{i\in\mathbb{Z}^+}$. We say that S is *syndetic* if there exists $d \in \mathbb{Z}^+$ such that $|s_{i+1} - s_i| \leq d$ for all $i \in \mathbb{Z}^+$. We say that S is *piecewise syndetic* if there exists $d \in \mathbb{Z}^+$ such that for any $n \in \mathbb{Z}^+$, there exists $j \in \mathbb{Z}^+$ such that $|s_{i+1} - s_i| \leq d$ for $i = j, j+1, \ldots, j+n-1$.

A set is syndetic if it has bounded gaps. The definition for piecewise syndetic has a lot of quantifiers but means that the set has bounded gaps on arbitrarily long intervals. Clearly, syndetic sets are also piecewise syndetic.

If S is a piecewise syndetic set then S contains arbitrarily long arithmetic progressions. To see this, let d be as in Definition 2.72 and take $n = w(k; d+1)$. Let $s_j, s_{j+1}, \ldots, s_{j+n-1}$ be those elements of S satisfying the bounded gap condition. Consider the following coloring of $[s_j, s_{j+n-1}]$: color $i \in [s_j, s_{j+n-1}]$ with color c when $i \in S+c$ and c is the minimal such integer. From the bounded gap condition we know that $0 \leq c \leq d$, so this defines a $(d+1)$-coloring of an interval of length at least $w(k; d+1)$. By van der Waerden's Theorem, we have a monochromatic k-term arithmetic progression in $S + j$ for some $j \in \{0, 1, \ldots, d\}$. Since arithmetic progressions are translation invariant, this means that S itself must also contain a k-term arithmetic progression.

This is pretty straightforward, and if all sets with positive upper density were piecewise syndetic, we would have a proof of Theorem 2.70. Unfortunately for this line of attack, our hypothesis is not true.

Theorem 2.73. *There exist sets of positive upper density that are not piecewise syndetic.*

The above theorem holds by a result found in [19] where they provide such a set:

$$\mathbb{Z}^+ \setminus \left(\bigcup_{n=2}^{\infty} \bigcup_{i=1}^{\infty} [in^3 + 1, in^3 + n] \right).$$

Remark. To tie piecewise syndetic sets to the coloring arena, Brown's Lemma [29] (see also [129] for a proof) states that every finite coloring of $\mathbb{Z}^+$ admits a monochromatic piecewise syndetic set.

Just as we can deduce van der Waerden's Theorem from the Hales-Jewitt Theorem, Szemerédi's Theorem would follow from the appropriate density version of the Hales-Jewitt Theorem, which is the subject of the last section in this chapter.

2.5.3 Density Hales-Jewett Theorem

We start with the statement of the theorem in this subsection's title:

Theorem 2.74 (Density Hales-Jewett Theorem)**.** *Let $\epsilon > 0$ and let $k, r \in \mathbb{Z}^+$. Let $\mathcal{W}(m)$ be the set of all variable words of length m over the alphabet $\{1, 2, \ldots, k\} \cup \{x\}$. There exists an integer $N = N(\epsilon, k)$ such that for any $n \geq N$, every subset $S \subseteq [1, k]^n$ with $|S| > \epsilon k^n$ admits $w(x) \in \mathcal{W}(n)$ with $\{w(i) : i \in [1, k]\} \subseteq S$.*

This result was originally proved by Furstenberg and Katznelson [77] using ergodic theory (see Section 5.2 for an introduction to ergodic theory techniques). Over 30 years later a combinatorial proof was developed [160]. Needless to say, both proofs are very deep and are not appropriate for this book. For a good exposition of Theorem 2.74, visit the last chapter of [161] (written by Steger; see the Foreword of [161]).

We remark that the $k = 2$ case follows by viewing $x_1x_2 \cdots x_n \in [1,2]^n$ as the subset of $[1,n]$ containing i if and only if $x_i = 2$. Consider the lattice of $[1,2]^n$ partially ordered by inclusion of the associated subsets, i.e., an edge connects two subsets if one is a subset of the other. We are searching for $w(x) \in \mathcal{W}(n)$ such that $w(1)$ and $w(2)$ are both in S. This means that we are looking for two subsets connected by an edge.

Recall the concept of an antichain in a poset. An antichain is a set of elements in the lattice with no edge between any of them. A fundamental poset result is Sperner's Lemma, which states that every antichain of the subsets of $[1,n]$ under inclusion partial ordering has size at most

$$\binom{n}{\lfloor \frac{n}{2} \rfloor} = \frac{n!}{\lfloor \frac{n}{2} \rfloor^2} \approx \frac{\sqrt{2\pi n}\left(\frac{n}{e}\right)^n}{\pi n \left(\frac{n/2}{e}\right)^n} = \frac{2^n}{\sqrt{\frac{\pi}{2}n}} = o(2^n),$$

where we are using Stirling's approximation. Lastly, note that if our set S has $|S| \geq \epsilon \cdot 2^n$, then S cannot be an antichain since it contains more than $o(2^n)$ elements. Hence, S must contain two elements with an edge between them, which is what we want to show.

2.6 Exercises

2.1 Is it possible to partition $[1,8]$ into two parts so that no part contains two (distinct) integers and their sum? Can the same be done with $[1,9]$? Does the same part of the partition of $[1,9]$ also contain a 3-term arithmetic progression? If not, prove that the other part must contain a 3-term arithmetic progression and a solution to $x + y = z$.

2.2 Prove that every r-coloring of $\mathbb{Z}^+$ admits a monochromatic set of the form $\{a, a+b, c, a+bc\}$.

2.3 Let k, s be integers such that $k, s \geq 2$. Define $B(k,s)$ to be the least positive integer such that any 2-coloring of $[1, B(k,s)]$ – for which no s consecutive integers are of the same color – admits a monochromatic k-term arithmetic progression of each color. Prove that $B(k,s)$ exist for all $k, s \geq 2$?

2.4 Show that $\widehat{w}(3,1;2) = 17$ (see Corollary 2.2).

2.5 Prove that every red-blue-coloring of $\mathbb{Z}^+$ admits either arbitrarily long red arithmetic progressions or arbitrarily long blue intervals.

2.6 Prove that for $c \in \mathbb{Z}^+ \setminus \{1\}$ the equation $y = cx$ is not 2-regular.

2.7 Define χ to be the following 3-coloring of $\mathbb{Z}^+$. For $i \in \mathbb{Z}^+$ with $i = 2^j k$ and $2 \nmid k$, let $\chi(i)$ be j modulo 3. Prove that χ does not admit a monochromatic solution to $x + 2y = 4z$.

2.8 Let $p \in \mathbb{Z}^+$ be an odd prime. Prove that $x + y = pz$ is not $2(p-1)$-regular.

2.9 Prove that every 2-coloring of $[1, k^2 - k - 1]$ admits a monochromatic solution to $\sum_{i=1}^{k-1} x_i = x_k$. These numbers are called generalized Schur numbers.

2.10 Show that for every r-coloring of $\mathbb{Z}^+$ we are guaranteed monochromatic Schur triples and monochromatic 3-term arithmetic progressions of the same color.

2.11 Let $m, k, r \in \mathbb{Z}^+$. By considering only the coloring of $m\mathbb{Z}^+$ for any given coloring of $\mathbb{Z}^+$, it is easy to see that there exists a least positive integer $w'(k, m; r)$ such that any r-coloring of $[1, w'(k, m; r)]$ admits a monochromatic k-term arithmetic progression $a, a+d, \ldots, a+(k-1)d$ with $d \equiv 0 \pmod{m}$. Prove this via Rado's Theorem.

2.12 Call a set $D \subseteq \mathbb{Z}^+$ *r-happy* if for every r-coloring of $\mathbb{Z}^+$ there exists an arbitrarily long monochromatic arithmetic progression $a, a+d, a+2d, \ldots,$ with $id \in D$ for some $i \in \mathbb{Z}^+$. Prove: If R is not r-happy and S is not s-happy, then $R \cup S$ is not rs-happy.

2.13 Let $k, r, s \in \mathbb{Z}^+$. Prove that $w(k; r+s-1) \leq w(w(k; r); s)$.

2.14 Prove that every r-coloring of an infinitely long geometric progression admits arbitrarily long monochromatic geometric progressions.

2.15 Show, via computer, that every 2-coloring of $[1, 252]$ admits a monochromatic set of the form $\{a, b, a+b, ab\}$ with $a \neq b$. This is due to Graham (and appears at the end of a paper by Hindman [110]).

2.16 Consider the same problem as in Exercise 2.15, but allow $a = b$. What is the minimal $n \in \mathbb{Z}^+$ such that every 2-coloring of $[1, n]$ admits a monochromatic set of the form $\{a, b, a+b, ab\}$? (This can be done by hand.)

2.17 Prove Lemma 2.58.

2.18 Consider the finite version of Theorem 2.66 formulated as follows: Let $k, \ell, r \in \mathbb{Z}^+$. There exists a minimal positive integer $n = n(k, \ell; r)$ such that every r-coloring of $[1, n]$ admits sets B and C with $|B| = k$ and $|C| = \ell$ such that $FS(B) \cup FP(C)$ is monochromatic. Show that $n(2, 2; 2) = 33$ under the condition that we allow $1 \in C$ so that for $s \neq 1$ we allow $\{1, s\} = FP(\{1, s\})$.

2.19 Let $S \subseteq \mathbb{Z}^+$ be a syndetic set. Prove that there exists an infinite $T \subseteq \mathbb{Z}^+$ such that $FS(T) \subseteq S - n$ for some $n \in \mathbb{Z}^+$. Does the same property hold if S is piecewise syndetic?

2.20 Let $g \in \mathbb{Z}^+$. If $S = \{s_1 < s_2 < s_3 < \cdots\}$ is a syndetic set such that $s_{i+1} - s_i \le g$ for all $i \in \mathbb{Z}^+$ then we say that S is *g-syndetic*. Let $D \subseteq \mathbb{Z}^+$ have the property that any $(2r+1)$-syndetic set admits arbitrary long arithmetic progressions $a, a+d, a+2d, \ldots$ with $d \in D$. Show that every r-coloring of $\mathbb{Z}^+$ admits arbitrarily long monochromatic arithmetic progressions $b, b+c, b+2c, \ldots$ with $c \in D$.

2.21 True or false: For $\epsilon > 0$ and n sufficiently large, if $A \subseteq [1, n]$ with $|A| > \epsilon n$, then there exist $a, b \in A$ such that $\sqrt{a+b} \in A$.

2.22 Let $k \in \mathbb{Z}^+$. Show that

$$L_k = \lim_{n \to \infty} \frac{r_k(n)}{n}$$

exists by using the fact that $r_k(n)$ is a subadditive function, i.e., that

$$r_k(m+n) \le r_k(m) + r_k(n)$$

(which must first be proved), and appealing to Fekete's Lemma (Lemma 1.6).

2.23 Use Theorem 2.67 to prove the equivalence of Theorems 2.70 and 2.71.

2.24 Let $k, s \ge 2$. Define $w_2(k, s)$ to be the least positive integer such that any 2-coloring of $[1, w_2(k, s)]$ admits either a monochromatic k-term arithmetic progression or s consecutive integers of the same color. Prove that $w_2(k, s) \le w(k; s)$ so that we have existence. Let $\Gamma_s(k)$ be the minimal positive integer Γ such that any set $\{x_1 < x_2 < \cdots < x_\Gamma\} \subseteq \mathbb{Z}^+$ with $x_{i+1} - x_i \le s$ contains a k-term arithmetic progression. Prove that $\Gamma_s(k)$ exists. Finally, prove that $w_2(k, s) \le 2\Gamma_s(k) - 1$.

2.25 For $s, k \in \mathbb{Z}^+$, define $\Omega_s(k)$ to be the least positive integer n such that every set $X = \{x_1, x_2, \ldots, x_n\} \subseteq \mathbb{Z}^+$ with $x_i \in [(i-1)s, is-1]$ contains a k-term arithmetic progression in X. Prove that $\Omega_s(k)$ exists and show that $\Gamma_s(k) < (r_k(s)+1) \cdot \Omega_s(k)$, where $\Gamma_s(k)$ is defined in Exercise 2.24.

3

Graph Ramsey Theory

Now I'm crawling the wallpaper that's looking more like a roadmap to misery.

–Andy Partridge

Ramsey theory earns its name from Frank Ramsey and his seminal theorem although Erdős and Graham (among others) have done more to solidify its contents as a subject worthy of study. Ramsey's main topic in the article where his eponymous theorem appears concerns Hilbert and Ackermann's Decision Problem. This problem asks if there is an algorithm that can decide, for an arbitrary statement of first-order logic, whether or not the statement is universally valid (the answer is that this is undecidable; see Church [44] and Turing [202]). The original language used in Ramsey's paper [166] in not couched in graph-theoretic terms as graph theory is a fairly newly organized subdiscipline of mathematics. Although Euler worked with graphs in the 1700s, the first textbook on graph theory, written by König [122], was published 6 years after Ramsey's result (according to Tutte's commentary that appears in the 1986 translation/reprint of König's book [123]). As the subject solidified, it became clear that graphs provide the perfect vehicle for discussions concerning Ramsey's Theorem.

3.1 Complete Graphs

We start by reminding the reader of a few definitions about graphs.

Definition 3.1 (Graph, Hypergraph, Degree, Edge, Hyperedge)**.** Let V be a set, called the *vertices*, and let E be a subset of $\wp(V) \setminus \emptyset$ (the power set of V excluding the empty set). We say that $G = (V, E)$ is a *graph* if all elements of E are subsets of size 2. In this case, E is referred to as the set of *edges*. For any given vertex v, the number of edges containing v is called the *degree of*

v. We say that $G = (V, E)$ is a *hypergraph* if some element of E is a subset of more than 2 elements. In this case, E is called the set of *hyperedges*.

Definition 3.2 (Subgraph, Subhypergraph). Let $G = (V, E)$ be a graph. If $V' \subseteq V$ and $E' \subseteq E$ then $H = (V', E')$ is a *subgraph* of G. If G is a hypergraph, then we call H a *subhypergraph*.

We will start by considering graphs (hypergraphs are treated later in this chapter). We will assume that any arbitrary graph we consider is connected so that there exists a string of edges from any vertex to any other vertex. We also assume that our graphs are simple, meaning that at most one edge exists between any two vertices. We will start by considering one of the fundamental graphs: the complete graph.

Definition 3.3 (Complete graph). Let $n \in \mathbb{Z}^+$. The *complete graph on n vertices*, denoted K_n, is $G = (V, E)$ where $|V| = n$ and E consists of all subsets of V of size 2.

We typically identify the vertex set V of K_n with $\{1, 2, \ldots, n\}$ and the edge set E with $\{\{i, j\} : i, j \in [1, n]; i < j\}$. In other words, all $\binom{n}{2}$ possible edges are present in a complete graph.

We start with the classical version of Ramsey's Theorem, restricted to the 2-color situation.

Theorem 3.4. *Let $k, \ell \in \mathbb{Z}^+$. There exists a minimal positive integer $n = n(k, \ell)$ such that if each edge of K_n is assigned one of two colors, say red and blue, then K_n admits a complete subgraph on k vertices with all edges red or a complete subgraph on ℓ vertices with all edges blue.*

Before getting to the proof, let's consider the $k = \ell = 3$ case. We will refer to K_3 as a triangle and will show that $n(3, 3) \leq 6$. Consider any 2-coloring of the edges of K_6. Isolate one vertex; call it X. Then X is connected to the other 5 vertices with either a red or a blue edge. Let R be the set of vertices connected to X with a red edge; let B be the set of vertices connected to X with a blue edge. We know that $|R \cup B| = 5$ and that $R \cap B = \emptyset$. Hence, $|R \sqcup B| = |R| + |B| = 5$. From this we can conclude by the pigeonhole principle that either $|R| \geq 3$ or $|B| \geq 3$. Without loss of generality, we assume $|R| \geq 3$.

At this stage we have X connected to at least 3 vertices, call them A, B, and C, via red edges. Consider the edges between these latter 3 vertices. If any edge is red, say $\{A, B\}$, then we have a red triangle (on vertices A, B, and X). In other words, we have deduced a K_3 with all edges red. On the other hand, if no edge is red, then they are all blue and we can conclude that the triangle on vertices A, B, and C has all blue edges, i.e., we have a K_3 with all blue edges.

The proof for general k and ℓ follows the same idea as the case for $k = \ell = 3$. We will isolate one vertex and separate the other vertices according to the color of the edge to the isolated vertex.

Proof of Theorem 3.4. We use induction on $k + \ell$. Since K_1 has no edge, we will assume $k, \ell \geq 2$. Thus, the base case is $k + \ell = 4$ and we need only show that $n(2, 2)$ exists. We easily have $n(2, 2) = 2$ since regardless of the color of the edge between two vertices, we have a monochromatic K_2 (which has only one edge). Note that $n(k, 2) = k$ and $n(2, \ell) = \ell$ so we may assume, below, that $k \geq 3$ and $\ell \geq 3$.

We will show that

$$n(k, \ell) \leq n(k - 1, \ell) + n(k, \ell - 1).$$

By the inductive assumption, we assume that both quantities on the right-hand side are finite, and consequently, by showing this inequality, $n(k, \ell)$ will be finite.

Let

$$m = n(k - 1, \ell) + n(k, \ell - 1)$$

and consider any 2-coloring of the edges of K_m. We will deduce the existence of either a red K_k or a blue K_ℓ.

Isolate one vertex, say X. Let R be the set of vertices connected to X with a red edge; let B be the set of vertices connected to X with a blue edge. We have $|R \sqcup B| = |R| + |B| = m - 1$. From the pigeonhole principle one of the following holds:

$$\text{(I)} \quad |R| \geq n(k - 1, \ell) \qquad \text{or} \qquad \text{(II)} \quad |B| \geq n(k, \ell - 1).$$

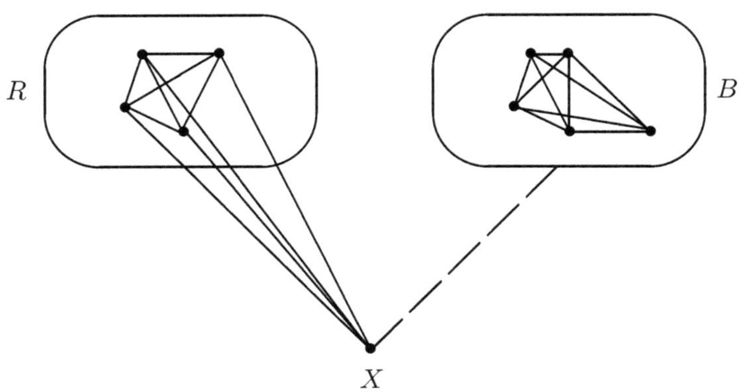

FIGURE 3.1
Visualizing the isolated vertex argument in the proof of Ramsey's Theorem

If (I) holds, then among the vertices in R we have either a blue K_ℓ and are done or we have a K_{k-1} subgraph on vertex set U with all edges red. If the latter holds, consider the K_k subgraph on vertex set $U \cup \{X\}$. Since all vertices in R are connected to X via a red edge, we see that this K_k subgraph has only red edges.

If (II) holds, then among the vertices in B we have either a red K_k and are done or we have a $K_{\ell-1}$ subgraph on vertex set W with all edges blue. Hence, the complete graph on vertex set $W \cup \{X\}$ is a blue K_ℓ. □

Definition 3.5 (2-color Ramsey number). The minimal numbers in Theorem 3.4 are called the 2-*color Ramsey numbers* and are denoted $R(k, \ell)$.

Ramsey numbers are notoriously difficult to determine. An exhaustive list of the known 2-color Ramsey numbers is given in Table 3.1.

TABLE 3.1
Known 2-color Ramsey numbers $R(k, \ell)$

		ℓ						
		3	4	5	6	7	8	9
k	3	6	9	14	18	23	28	36
	4	9	18	25				

Of course, upper and lower bounds on other numbers are known, but matching these bounds to determine the exact value is a difficult endeavor. Lower bounds are far easier as we only need to find one graph that avoids the proposed monochromatic structures. For upper bounds we must deduce that a monochromatic structure is contained in every possible 2-coloring. Brute force checking quickly becomes infeasible since there are $2^{\binom{n}{2}}$ ways to 2-color the edges of K_n. So, for example, checking all 2-colorings of the edges of K_{25} means checking 2^{300} graphs (a number which is roughly 200 million times the approximated number of atoms in the observable universe). In terms of computing time, the world's (current) fastest computer (named Summit) can perform roughtly 2^{60} operations per second, or about 2^{85} operations per year.

From the proof of Theorem 3.4 we have

$$R(k, \ell) \leq R(k-1, \ell) + R(k, \ell-1)$$

so that we have a way to determine reasonable upper bounds. These are typically not best possible. In fact, we can improve on this bound slightly when both $R(k-1, \ell)$ and $R(k, \ell-1)$ are even: $R(k, \ell) \leq R(k-1, \ell) + R(k, \ell-1) - 1$.

To see this, we follow the proof, and use the notation, of Theorem 3.4. Letting

$$m = R(k-1, \ell) + R(k, \ell-1) - 1,$$

we may assume that $|R| = R(k-1, \ell) - 1$ and $|B| = R(k, \ell-1) - 1$; otherwise, one of (I) or (II) above Figure 3.1 is true and the result follows. The key observation is that these values of $|R|$ and $|B|$ must be true regardless of which vertex X we isolate. Thus, we conclude that every vertex has exactly

$R(k-1,\ell)-1$ red edges. Counting red edges by summing over vertices and noting that doing so counts each edge twice (once for each vertex of an edge), we find that the total number of red edges $|E_R|$ in K_m must be

$$2|E_R| = \sum_{v=1}^{m} (R(k-1,\ell)-1) = m\,(R(k-1,\ell)-1)\,.$$

Since m is an odd integer and $R(k-1,\ell)$ is even (this is where we need $R(k-1,\ell)$ and $R(k,\ell-1)$ to both be even), we see that $m\,(R(k-1,\ell)-1)$ is an odd integer equal to $2|E_R|$, which is a contradiction.

The best-known asymptotic bounds on the 2-color diagonal Ramsey numbers are

$$\frac{k}{e}\left(\sqrt{2}\right)^{k+1}(1+o(1)) \leq R(k,k) \leq \frac{4^{k-1}}{(k-1)^{c\log k/\log\log k}}(1+o(1)), \qquad (3.1)$$

for some constant $c>0$, where the lower bound is due to Spencer [194] and the upper bound is due to Conlon [49]. We will derive the lower bound in Section 6.1. We can quickly give an upper bound close to the one above. More generally, we claim that

$$R(k,\ell) \leq \binom{k+\ell-2}{k-1}.$$

It is easy to check that this holds when at least one of k and ℓ equals 2. Hence, we may induct on $k+\ell$. Following the proof of Ramsey's Theorem and applying our inductive hypothesis, we have

$$\begin{aligned} R(k,\ell) &\leq R(k,\ell-1)+R(k-1,\ell) \\ &\leq \binom{k+\ell-3}{k-1}+\binom{k+\ell-3}{k-2} \\ &= \binom{k+\ell-2}{k-1}, \end{aligned}$$

which is what we wanted to show. Now, using Stirling's approximation with $k=\ell$, we have (for large k):

$$\begin{aligned} R(k,k) \leq \binom{2(k-1)}{k-1} = \frac{(2(k-1))!}{((k-1)!)^2} &\approx \frac{2\sqrt{\pi(k-1)}\left(\frac{2(k-1)}{e}\right)^{2(k-1)}}{2\pi(k-1)\left(\frac{(k-1)}{e}\right)^{2(k-1)}} \\ &= \frac{4^{k-1}}{\sqrt{\pi(k-1)}}. \end{aligned}$$

For further information about Ramsey numbers, the definitive resource is the dynamic survey by Radziszowski [165].

As Ramsey theory's creed is that structures survive under any partitioning – not just into two parts – we now state Ramsey's Theorem for an arbitrary number of colors.

Theorem 3.6 (Ramsey's Theorem). *Let $r \in \mathbb{Z}^+$ and let $k_1, k_2, \ldots, k_r \in \mathbb{Z}^+$. There exists a minimal positive integer $n = n(k_1, k_2, \ldots, k_r)$ such that every r-coloring of the edges of K_n with colors $1, 2, \ldots, r$ contains, for some $j \in \{1, 2, \ldots, r\}$, a K_{k_j} subgraph with all edges of color j.*

Proof. The proof is an easy extension of the proof of Theorem 3.4. Formally, we are inducting on r (we have done $r = 2$ in Theorem 3.4) and then using induction on $\sum_{i=1}^{r} k_i$ with a base case of $\sum_{i=1}^{r} k_i = 2r$. Again, we have $n(2, 2, \ldots, 2) = 2$ since regardless of the color of the edge in K_2, we have a monochromatic K_2.

First, consider the situation when $k_i = 2$ for some $i \in \{1, 2, \ldots, r\}$. Then $n(k_1, k_2, \ldots, k_{i-1}, 2, k_{i+1}, \ldots, k_r) = n(k_1, k_2, \ldots, k_{i-1}, k_{i+1}, \ldots, k_r)$ as any use of color i gives a monochromatic K_2. From this equality, since the latter expression has $r - 1$ colors, by induction on r we are done in the case of some k_i equaling 2. Hence, we may now assume that $k_i \geq 3$ for all $i \in \{1, 2, \ldots, r\}$.

We proceed by showing that

$$n(k_1, k_2, \ldots, k_2) \leq \sum_{i=1}^{r} n(k_1, k_2, \ldots, k_{i-1}, k_i - 1, k_{i+1}, \ldots, k_r).$$

Toward this end, let

$$m = \sum_{i=1}^{r} n(k_1, k_2, \ldots, k_{i-1}, k_i - 1, k_{i+1}, \ldots, k_r)$$

and consider an arbitrary r-coloring of the edges of K_m.

Isolate a vertex X in K_m and partition the remaining vertices by letting C_i, for $i \in \{1, 2, \ldots, r\}$, be the set of vertices connected to X by an edge of color i. Then for some $j \in \{1, 2 \ldots, r\}$ we have

$$|C_j| \geq n(k_1, k_2, \ldots, k_{j-1}, k_j - 1, k_{j+1}, \ldots, k_r)$$

for otherwise we would conclude that

$$\begin{aligned} m - 1 = \sum_{i=1}^{r} |C_i| &\leq \sum_{i=1}^{r} \left(n(k_1, k_2, \ldots, k_{i-1}, k_i - 1, k_{i+1}, \ldots, k_r) - 1 \right) \\ &= m - r, \end{aligned}$$

which is, of course, untrue for $r \geq 2$ (the $r = 1$ case of the theorem is trivial). So, we now have a complete graph on at least $n(k_1, k_2, \ldots, k_{j-1}, k_j -$

$1, k_{j+1}, \ldots, k_r)$ vertices, all of which are connected to X by an edge of color j. By the inductive assumption on $\sum_{i=1}^{r} k_i$ we have two cases to consider. First, if we have a monochromatic K_{k_i} of color i for some $i \in \{1, 2, \ldots, r\} \setminus \{j\}$ then we are done. If, on the other hand, we have a monochromatic $K_{k_j - 1}$ of color j, say on the vertices U, then the complete graph on the vertices $U \cup \{X\}$ is a complete graph on k_j vertices with all edges of color j and, again, we are done. □

The notation for these r-color Ramsey numbers is $R(k_1, k_2, \ldots, k_r)$. In the situation where all k_i's are equal, we write $R(k; r)$. The only known values for these multicolored Ramsey numbers are $R(3, 3, 3) = R(3; 3) = 17$ and $R(3, 3, 4) = 30$, the latter value being the most recently determined Ramsey number [47].

3.1.1 Deducing Schur's Theorem

As our first "application" of Ramsey's Theorem, we will deduce Schur's Theorem. Recall that Schur's Theorem states that there exists a minimal positive integer $s(r)$ such that every r-coloring of $[1, s(r)]$ admits a monochromatic solution to $x + y = z$. So, we need to deduce integer solutions from a graph. We do so by considering a subclass of colorings of K_n.

Definition 3.7 (Difference coloring). A *difference coloring* of the edges of K_n is one in which the color of every edge $\{i, j\}$ depends solely on $|i - j|$.

With this definition, if $\chi : [1, n-1] \to \{0, 1, \ldots, r-1\}$ is a coloring of integers, then we have the induced difference coloring of K_n where we color edge $\{i, j\}$ with $\chi(|i - j|)$.

To deduce Schur's Theorem, let $n = R(3; r)$. For any r-coloring of $[1, n-1]$ consider the difference coloring of the edges of K_n. By Ramsey's Theorem, this difference coloring admits a monochromatic K_3, say on the vertices u, v, and w, with $u < v < w$. This means that the integers $v - u, w - v$, and $w - u$ all have the same color. Let $x = v - u$, $y = w - v$, and $z = w - u$ to see that we have a solution to $x + y = z$ with x, y, and z all the same color. Consequently, we see that $s(r) \le R(3; r) - 1$.

The above argument can be easily extended to show that any r-coloring of the integer interval $[1, R(k; r) - 1]$ admits a monochromatic solution to $\sum_{i=1}^{k-1} x_i = x_k$. The minimal positive integer $n = n(k; r)$ such that every r-coloring of $[1, n]$ admits a monochromatic solution to $\sum_{i=1}^{k-1} x_i = x_k$ is referred to as a *generalized Schur number*. This argument gives $n(k; r) \le R(k; r) - 1$. It is known [21] that

$$n(k; 2) = k^2 - k - 1;$$

however, $n(k; r)$ for $r \ge 3$ is unknown: the bound

$$n(k; r) \ge \frac{k^{r+1} - 2k^r + 1}{k - 1}$$

has been determined [21]. As we can see, the bound between these two Ramsey-type numbers is quite weak since we have seen that $(\sqrt{2})^k < R(k,k)$ (see Inequality (3.1)) so that the $k^2 - k$ lower bound on $R(k,k)$ is not strong. The reason for this is because we do not use the whole complete monochromatic subgraph in deducing the existence of $n(k;r)$.

3.2 Other Graphs

As stated at the end of the last section, the proof that generalized Schur numbers exist as a consequence of Ramsey's Theorem does not use the full power of Ramsey's Theorem. In particular, we do not need a monochromatic complete graph; we only need the "outside" edges of this complete graph to have the same color, i.e., a monochromatic cycle.

Definition 3.8 (Path, Cycle, Tree). Let $G = (V, E)$ be the graph with vertex set $V = \{v_1, v_2, \ldots, v_n\}$ and edge set $E = \{\{v_i, v_{i+1}\} : i \in \{1, 2, \ldots, n-1\}\}$. Then G is called a *path* (from v_1 to v_n, or v_n to v_1) and we denote it by P_n. If we add the edge $\{v_1, v_n\}$ to the edge set, then G is called an *n-cycle* and we denote it by C_n. Paths are a subclass of the class of graphs known as *trees*. A tree T_n is a graph on n vertices with no k-cycle as a subgraph, for any k.

There are many other named graphs (for example, in Figure 3.2 we present a graph attributed to Kempe but commonly referred to as the Peterson graph, which is useful as it often serves as a counterexample). Complete graphs, paths, cycles, and trees are some of the important ones. Below we define two more important classes of graphs.

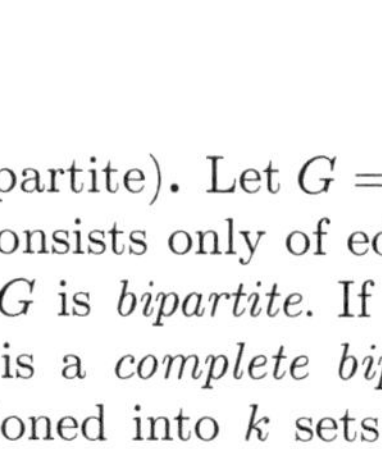

FIGURE 3.2
The Peterson graph

Definition 3.9 (Bipartite, k-partite). Let $G = (V, E)$. If V can be partitioned into V_1 and V_2 such that E consists only of edges with one vertex in each of V_1 and V_2, then we say that G is *bipartite*. If every edge between V_1 and V_2 is in E then we say that G is a *complete bipartite graph* and denote it by $K_{|V_1|,|V_2|}$. If V can be partitioned into k sets $V_1, V_2, \ldots, V_k$ such that every edge contains at most one vertex from each V_i then we say that G is k-partite. If every possible edge between every pair V_i and V_j, $i \neq j$, is in E then we say that G is a *complete k-partite graph.*

To aid in Definition 3.9, consider Figure 3.3, below. It is worth noting that, although we consider arbitrary graphs to be connected, many k-partite graphs are not connected; however, complete k-partite graphs are connected.

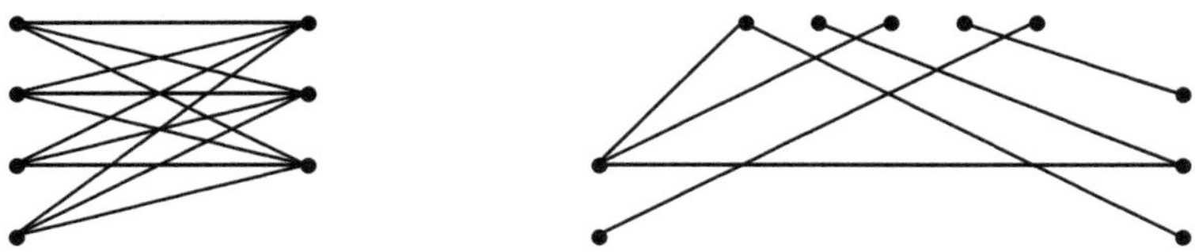

(a) $K_{4,3}$ (b) a 3-partite graph

FIGURE 3.3
Examples of bipartite and k-partite graphs

Because every type of graph exists as a subgraph of some sufficiently-sized complete graph, we can make the following definition.

Definition 3.10 (Graph Ramsey number)**.** Let $r \in \mathbb{Z}^+$ and let $G_1, G_2, \ldots, G_r$ be graphs. The minimal positive integer $n = R(G_1, G_2, \ldots, G_r)$ such that every r-coloring (with colors $1, 2, \ldots, r$) of the edges of K_n admits an edge-wise monochromatic subgraph G_j of color j is called a *graph Ramsey number*. In the situation when $G = G_1 = G_2 = \cdots = G_r$, we write $n = R(G; r)$.

To formally prove the existence of graph Ramsey numbers, let G_i be a graph on k_i vertices. Since G_i is a subgraph of K_{k_i} we have $R(G_1, G_2, \ldots, G_r) \leq R(k_1, k_2, \ldots, k_r)$. In fact, one reason for studying graph Ramsey numbers is because of this connection to Ramsey numbers.

3.2.1 Some Graph Theory Concepts

Because we are now in the situation where graphs are no longer of a fixed form, we define some basic statistics on them to aid in our investigation.

Definition 3.11 (Clique number)**.** Let G be a graph. The *clique number of* G, denoted $\omega(G)$, is the maximal number of vertices n in G such that K_n is a subgraph of G.

Definition 3.12 (Chromatic number of a graph)**.** Let G be a graph. The *chromatic number of* G, denoted $\chi(G)$, is the minimal number of colors r such that we can color the vertices of G in such a way that no two vertices connected by an edge are assigned the same color.

Note that in Definition 3.12 we have moved from coloring edges to coloring vertices. To aid in discussions of the chromatic number, we make the following definition.

Definition 3.13 (Vertex-valid, c-colorable). We call a coloring of the vertices of a graph a *vertex-valid coloring* if no two vertices connected by an edge are assigned the same color. If it is possible to color the vertices of a graph with c colors so that no two vertices connected by an edge are assigned the same color, we say that the graph is *c-colorable.*

The last graph statistic we will use is next.

Definition 3.14 (Independence number of a graph). Let G be a graph. The *independence number of* G, denoted $\alpha(G)$, is the maximal number of vertices of G such that no two of them are connected by an edge (i.e., G contains a subgraph on $\alpha(G)$ vertices with no edge).

The above three statistics allow us to gain some information about arbitrary graphs; however, in general, they are difficult to determine. We do have the following relationship between the three:

$$\chi(G) \geq \max\left\{\omega(G), \left\lceil \frac{|G|}{\alpha(G)} \right\rceil\right\},$$

where $|G|$ denotes the number of vertices of G.

Clearly, $\chi(G) \geq \omega(G)$ since the complete graph on $\omega(G)$ vertices requires $\omega(G)$ colors to be vertex-valid (since every pair of vertices is connected by an edge). To see that $\chi(G) \geq \left\lceil \frac{|G|}{\alpha(G)} \right\rceil$, note that any subgraph on $\alpha(G)+1$ vertices must use at least 2 colors to be a vertex-valid coloring. Hence, at most $\alpha(G)$ vertices may use the same color. We deduce that we must use at least $\left\lceil \frac{|G|}{\alpha(G)} \right\rceil$ colors to produce a vertex-valid coloring of G; otherwise, some set of $\alpha(G)+1$ vertices uses the same color.

In the next subsection, we will use the above statistics to give some results on graph Ramsey numbers.

3.2.2 Graph Ramsey Numbers

The first result we present holds definitionally and is useful for translating between results couched in typical Ramsey theory language and results given in standard graph theory language. Letting G be any graph, we interpret it as a 2-coloring of the edges of $K_{|G|}$ by coloring all edges in G red and the remaining edges (not present in G) blue. This coloring has no red $K_{\omega(G)+1}$ and no blue $K_{\alpha(G)+1}$ by definition. This yields the following result.

Lemma 3.15. *Let G be any graph. Then $R(\omega(G)+1, \alpha(G)+1) \geq |G|+1$.*

Applying Lemma 3.15 with G being a 5-cycle we easily have $\omega(G) = 2$ and $\alpha(G) = 2$ so that $R(3,3) \geq 6$, which as we have seen is tight.

The next general result is a special case of a result of Chvátal and Harary [46]. We remind the reader that all arbitrary graphs we consider are connected.

Theorem 3.16. *Let G and H be graphs. Then*

$$R(G,H) \geq (\chi(G) - 1)(|H| - 1) + 1.$$

Proof. Let

$$n = (\chi(G) - 1)(|H| - 1).$$

We will exhibit a 2-coloring of the edges of K_n with no red G and no blue H, from which the inequality follows. Partition the vertices of K_n into $\chi(G) - 1$ parts of $|H| - 1$ vertices each. In each part, color all edges between the $|H| - 1$ vertices blue. Color the remaining edges of K_n red. Hence, every edge between copies of $K_{|H|-1}$ is red. Clearly this coloring admits no blue H as H cannot be a subgraph of $K_{|H|-1}$. We must show that there is no red G.

Assume, for a contradiction, that this coloring admits a red G. Then G may have at most one vertex in each copy of $K_{|H|-1}$. Hence, G can have at most $\chi(G)-1$ vertices. However, this means we can use $\chi(G)-1$ colors to color the vertices of G to produce a vertex-valid coloring of G. This contradicts the definition of $\chi(G)$ as being the minimal such number. □

Applying Theorem 3.16, we give our first result for specific types of graphs. This result is due to Chvátal [45]. Recall that T_m is a tree on m vertices.

Corollary 3.17. *Let $m, n \in \mathbb{Z}^+$. For any given tree T_m, we have*

$$R(T_m, K_n) = (m - 1)(n - 1) + 1.$$

Proof. Using $G = K_n$ and $H = T_m$ in Theorem 3.16 we immediately have $R(T_m, K_n) \geq (m - 1)(n - 1) + 1$ so it remains to show that $R(T_m, K_n) \leq (m-1)(n-1)+1$. To show this we induct on $m+n$, with $R(T_2, K_n) = n$ and $R(T_m, K_2) = m$ being trivial. Note that the inductive assumption means the formula holds for any type of tree on less than m vertices.

Let K be the complete graph on vertices V with

$$|V| = (m - 1)(n - 1) + 1.$$

Assume, for a contradiction, that some 2-coloring of the edges of K avoids both a blue K_n and a red T_m for some type of tree T_m. By the inductive assumption, any complete subgraph on $(m-2)(n-1)+1$ vertices has either a red tree of any specified type on $m-1$ vertices or a blue K_n. In the latter case, we are done, so we may assume that we have a red tree S on the vertices M, with $|M| = m-1$. Furthermore, we choose S to be a tree obtainable from T_m by the removal of one vertex. Let the removed vertex be attached to vertex u

in S. Then every edge from u to $v \in V \setminus M$ must be blue, else we have a red T_m and are done.

Consider the complete subgraph on the vertices $V \setminus M$. This subgraph has $(m-1)(n-2)+1$ vertices. By the inductive assumption, we have either a red T_m, and are done, or we have a blue K_{n-1} on the vertices $W \subseteq V \setminus M$. Going back to K, we see that the complete subgraph on $W \cup \{u\}$ is a blue K_n, and we are done. □

Sticking with trees, a next obvious question is: What can we say about the value of $R(T_m; 2)$? This turns out to be quite a difficult problem. Recently, Ajtai, Komlós, Simonovits, and Szemerédi (see, e.g., [2]), announced a proof that every graph on n vertices having more than $\frac{(m-2)}{2}n$ edges contains every type of tree on m vertices, provided m is sufficiently large. This result settles (mostly) a more than 50-year-old conjecture of Erdős and Sós. From this we can conclude that for m sufficiently large,

$$R(T_m; 2) \leq \begin{cases} 2m-2 & \text{for } m \text{ even;} \\ 2m-3 & \text{for } m \text{ odd.} \end{cases}$$

To see this, first consider m even. Then K_{2m-2} contains $\binom{2m-2}{2}$ edges, of which at least half must be of one color, say red. Hence, viewing only the red edges, if

$$\frac{1}{2}\binom{2m-2}{2} > \frac{(m-2)(2m-2)}{2}$$

then we have every type of tree on m vertices, with all edges red. Since this inequality holds, we are done. The case for m odd is similar with the additional observation that $\frac{1}{2}\binom{2m-3}{2}$ is not an integer.

We end this section with a selection of formulas for various pairings of graphs and point the reader to [31, 32, 84, 163] for further discussions on, and values and bounds on, graph Ramsey numbers.

Theorem 3.18. *Let $k, \ell \in \mathbb{Z}^+$. Then the following hold:*

(i)

$$R(K_{1,k}, K_{1,\ell}) = \begin{cases} k+\ell-1 & \textit{if } k \textit{ and } \ell \textit{ are both even;} \\ k+\ell & \textit{otherwise;} \end{cases}$$

(ii) for $k \geq \ell \geq 3$:

$$R(C_k, C_\ell) = \begin{cases} 2k-1 & \ell \textit{ odd and } (k,\ell) \neq (3,3); \\ k+\frac{\ell}{2}-1 & k,\ell \textit{ both even, } (k,\ell) \neq (4,4); \\ \max\left\{2k-1, k+\frac{\ell}{2}-1\right\} & k \textit{ odd and } \ell \textit{ even;} \end{cases}$$

(iii) $R(P_k, P_\ell) = k + \left\lfloor \frac{\ell}{2} \right\rfloor - 1$ for $k \geq \ell \geq 2$.

The first formula is from [99]; the second is from [74] and, independently, [172]; the last is from [80].

The astute reader may wonder why the formula for $R(C_k, C_k)$ is less than $k^2 - k - 1$ (the value of the generalized Schur number) since C_k is the "outside" of K_k. This is because we may have a cycle on k vertices without having K_k. Consider Figure 3.4. We see that the outside edges of K_4 are not all the same color, yet we do have a monochromatic C_4 with $1 \to 3 \to 4 \to 2 \to 1$.

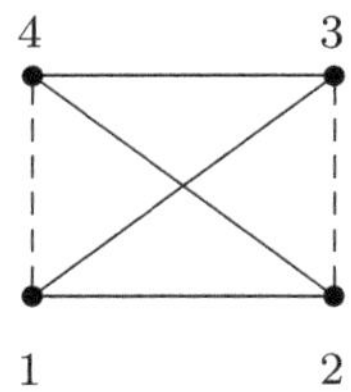

FIGURE 3.4
A monochromatic C_4 without a monochromatic K_4

3.3 Hypergraphs

Consider a standard 6-sided die, sketched with its corners labeled as in Figure 3.5.

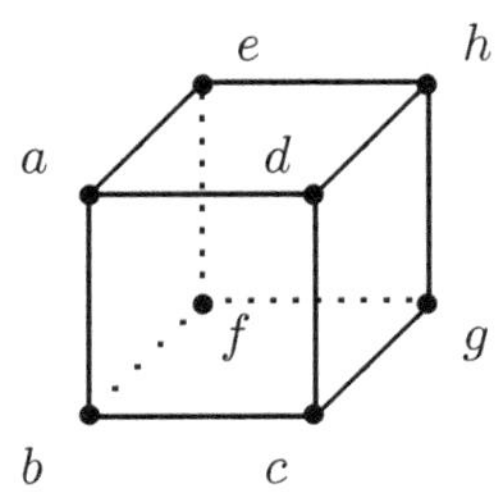

FIGURE 3.5
Moving from graphs to hypergraphs

Describing the die in Figure 3.5 as a graph on 8 vertices, our edge set would be

$$\{\{a,b\},\{a,d\},\{a,e\},\{b,c\},\{b,f\},\{c,d\},\{c,g\},\{d,h\},\{e,f\},\{e,h\},\{f,g\},\{g,h\}\}.$$

But this is not how we would normally describe a standard die. We typically note the 6 sides of a die and not the edges as we have given them. We may call the sides faces; however, in order to abstract the idea of a graph, let's call them some type of edge. For the same reason we abstract a plane in 3 dimensions to a hyperplane in more than 3 dimensions, we abstract edges of more than 2 vertices to hyperedges.

Recalling Definition 3.1, we may refer to the faces of our cube as hyperedges and describe the associated hypergraph as being on the vertices a, b, c, d, e, f, g, h with hyperedge set

$$\{\{a, b, c, d\}, \{a, b, e, f\}, \{a, d, e, h\}, \{b, c, f, g\}, \{c, d, g, h\}, \{e, f, g, h\}\}.$$

The reader may notice that each hyperedge in this description contains exactly 4 vertices. This is an example of a hypergraph in the class a hypergraphs with which we will be dealing.

Definition 3.19 (ℓ-uniform hypergraph)**.** Let $G = (V, E)$ be a hypergraph. Let $\ell \in \mathbb{Z}^+$ with $\ell \geq 3$. If every element of E contains exactly ℓ vertices from V, then we say that G is an *ℓ-uniform hypergraph.* If E contains all $\binom{|V|}{\ell}$ subsets of V of size ℓ, then we call G the *complete ℓ-uniform hypergraph* and denote it by K_n^ℓ, where $n = |V|$.

Applying this definition to the die example above, our second description is one of a 4-uniform hypergraph; however, it is not a complete 4-uniform hypergraph since it has only 6 hyperedges and not all $\binom{6}{4} = 15$ hyperedges that a K_6^4 has.

3.3.1 Hypergraph Ramsey Theorem

In Ramsey's Theorem we color edges and deduce monochromatic complete subgraphs. Replacing edges with hyperedges and subgraphs with subhypergraphs is the basis of the Hypergraph Ramsey Theorem, which is actually the original theorem proved by Ramsey in his seminal paper [166, Theorem A].

Formally, for a hypergraph $H = (V, E)$, each hyperedge in E is assigned a color and a subhypergraph on vertices $U \subseteq V$ may only have a hyperedge $e \in E$ provided $e \in \wp(U)$.

Notation. We will refer to a hyperedge consisting of ℓ vertices as an *ℓ-hyperedge.*

Theorem 3.20 (Hypergraph Ramsey Theorem)**.** *Let $\ell, r \in \mathbb{Z}^+$ with both at least 2. For $i \in \{1, 2, \ldots, r\}$, let $k_i \in \mathbb{Z}^+$ with $k_i \geq \ell$. Then there exists a minimal positive integer $n = R_\ell(k_1, k_2, \ldots, k_r)$ such that every r-coloring of the ℓ-hyperedges of K_n^ℓ with the colors $1, 2, \ldots, r$ contains, for some $j \in \{1, 2, \ldots, r\}$, a $K_{k_j}^\ell$ subhypergraph with all ℓ-hyperedges of color j.*

Proof. We start by proving the $r = 2$ case. We prove this via induction on ℓ. The $\ell = 2$ case of this theorem is Ramsey's Theorem (Theorem 3.6), so that we have already proved the base case. Given k_1 and k_2, by the inductive assumption we assume that $n = R_{\ell-1}(k_1, k_2)$ exists. For ease of exposition, we will use the colors red and blue, with red associated with k_1 (and blue with k_2).

Inside of the induction on ℓ, we induct on $k_1 + k_2$, as we did in the proof of Ramsey's Theorem. So, in our pursuit of showing that $R_\ell(k_1, k_2)$ exists, we may assume that $R_\ell(k_1 - 1, k_2)$ and $R_\ell(k_1, k_2 - 1)$ both exist. The base cases for our induction on $k_1 + k_2$ are trivial: $R_\ell(\ell, k_2) = k_2$ and $R_\ell(k_1, \ell) = k_1$ (for the first, we either have a red hyperedge and, hence, a red K_ℓ^ℓ, or all hyperedges are blue and we have a blue $K_{k_2}^\ell$; for the second, reverse the colors and use k_1 instead of k_2). Hence, we may assume that k_1 and k_2 are both greater than ℓ. Let

$$n = R_{\ell-1}(R_\ell(k_1 - 1, k_2), R_\ell(k_1, k_2 - 1)) + 1.$$

We will show that $R_\ell(k_1, k_2) \leq n$.

Consider an arbitrary 2-coloring χ of the ℓ-hyperedges of K_n^ℓ with vertex set V. Isolate a vertex $v \in V$. Consider the complete $(\ell - 1)$-uniform hypergraph on vertices $V \setminus \{v\}$ where the coloring $\widehat{\chi}$ of the $(\ell - 1)$-hyperedges is inherited from χ in the following way: For e an $(\ell - 1)$-hyperedge, let

$$\widehat{\chi}(e) = \chi(e \cup \{v\}).$$

From the definition of n we have either a complete $(\ell - 1)$-uniform hypergraph on $R_\ell(k_1 - 1, k_2)$ vertices with all $(\ell - 1)$-hyperedges red under $\widehat{\chi}$ or a complete $(\ell - 1)$-uniform hypergraph on $R_\ell(k_1, k_2 - 1)$ vertices with all $(\ell - 1)$-hyperedges blue under $\widehat{\chi}$. Without loss of generality, we may assume the latter holds.

Let W be the set of $R_\ell(k_1, k_2 - 1)$ vertices with all $(\ell - 1)$-hyperedges blue under $\widehat{\chi}$. By the assumed existence of $R_\ell(k_1, k_2 - 1)$ we have, under χ, either a red $K_{k_1}^\ell$ and are done, or we have a blue $K_{k_2-1}^\ell$ on vertex set U with the property that all $(\ell - 1)$ hyperedges are blue under $\widehat{\chi}$. Noting that $v \notin U$, consider the complete ℓ-uniform hypergraph on the vertices $U \cup \{v\}$. The ℓ-hyperedges that include v are all blue since, by the definition of $\widehat{\chi}$, we have $\widehat{\chi}(e) = \chi(e \cup \{v\})$ for any $(\ell - 1)$-hyperedge e with vertices in $V \setminus \{v\}$, which we have deduced are all blue in $K_{k_2-1}^\ell$. Thus, all ℓ-hyperedges on $U \cup \{v\}$ are blue and we have a blue $K_{k_2}^\ell$. This completes the $r = 2$ case of the theorem.

We now move on to proving the theorem for an arbitrary number of colors, with $r = 2$ serving as the base case of induction on r. We will show that

$$R_\ell(k_1, k_2, \ldots, k_r) \leq R_\ell(R_\ell(k_1, k_2), k_3, k_4, \ldots, k_r),$$

where the upper bound exists by the inductive assumption (there are only $r-1$ colors). To see this, consider an arbitrary r-coloring χ of the ℓ-hyperedges of K_n^ℓ, where

$$n = R_\ell(R_\ell(k_1, k_2), k_3, k_4, \ldots, k_r).$$

For our exposition, to show that $R_\ell(k_1, k_2, \ldots, k_r)$ exists, let k_1 be associated with the color dark blue and k_2 with the color light blue.

We next define an $(r-1)$-coloring of the ℓ-hyperedges of K_n^ℓ by identifying dark blue and light blue; in other words, these colors are now just blue. By the inductive assumption, we either have a $K_{k_j}^\ell$ with all ℓ-hyperedges of color j for some $j \in \{3, 4, \ldots, r\}$ (i.e., j is not blue), and are done, or we have a blue complete ℓ-uniform hypergraph on $R_\ell(k_1, k_2)$ vertices. We now return to the original shades of blue so that we have a 2-coloring of the ℓ-hyperedges of the complete graph on $R_\ell(k_1, k_2)$ vertices. Since we are now in the $r = 2$ case, which we proved above, we have either a dark blue $K_{k_1}^\ell$ or a light blue $K_{k_2}^\ell$, and are done. □

Given the difficulty of determining Ramsey numbers, it should come as little surprise that we do not know the values of many (non-trivial) hypergraph Ramsey numbers. In fact, we only know the value of one such number: $R_3(4, 4) = 13$, which was determined in 1991 [144]. Furthermore, the general bounds for $R_\ell(k, k)$ are very large and not easily developed; see [51] for current information.

3.3.2 Deducing Arnautov-Folkman-Rado-Sanders' Theorem

Recall Theorem 2.55 about monochromatic sumsets:

Theorem 2.55 (Arnautov-Folkman-Rado-Sanders' Theorem). *Let $k, r \in \mathbb{Z}^+$. There exists a minimal positive integer $n = n(k; r)$ such that every r-coloring of $[1, n]$ admits $S \subseteq [1, n]$ with $|S| = k$ such that $FS(S)$ is monochromatic.*

As mentioned in the last section, we did not use the full power of Ramsey's Theorem to prove the existence of monochromatic solutions to $\sum_{i=1}^{k-1} x_i = x_k$ as we only used the "outside" cycle of a complete graph. By looking at the other edges we see that we have many elements of $FS(x_1, x_2, \ldots, x_{k-1})$ of the same color. However, we do not have all of them; we only have sums of the form $x_j + x_{j+1} + \cdots + x_{j+c}$, i.e., sums with consecutive indices. A further complication arises when we notice that we do not know whether or not all of the x_i's are distinct. Hence, in order to use Ramsey's Theorem (Theorem 3.6), we must take a different approach, and, perhaps, Theorem 3.6 may not be strong enough to deduce Theorem 2.55 (and you would be correct in assuming that it probably isn't, based on the section in which this material resides).

We will follow Sander's proof [178] of Theorem 2.55, which uses the following interesting consequence of the Hypergraph Ramsey Theorem.

Lemma 3.21. *Let $k, t, r \in \mathbb{Z}^+$ with $k \geq t$. Then there exists an integer $n = N(k, t; r)$ such that for any set B with $|B| \geq n$ and any r-coloring of all non-empty subsets of B, there exists $A \subseteq B$ with $|A| = k$ and a sequence of t (not necessarily distinct) colors $i_1, i_2, \ldots, i_t$ such that every subset of A of size j has color i_j, for $1 \leq j \leq t$.*

Proof. We induct on t. For $t = 1$ we may take $N(k, 1; r) = (k-1)r + 1$ since there are $(k-1)r+1$ subsets of size 1 each being assigned one of r colors, from which we can deduce that one color is used at least k times. Let A be the set of any k of these elements.

We now assume that $N(k,t-1;r)$ exists for all k and r and will prove that $N(k,t;r)$ exists for all k and r. We will show that

$$N(k,t;r) \leq N(R_t(k;r), t-1; r),$$

where $R_t(k;r)$ is the r-color hypergraph Ramsey number.

Let B be a set with $N(R_t(k;r), t-1; r)$ elements. By the inductive assumption, for any r-coloring of the sets in $\wp(B) \setminus \emptyset$ there exists $C \subseteq B$ of size $R_t(k;r)$ and a sequence of $t-1$ colors $i_1, i_2, \ldots, i_{t-1}$ such that every subset of C of size j has color i_j for $1 \leq j \leq t-1$. By the Hypergraph Ramsey Theorem, $R_t(k;r)$ gives us a monochromatic K_k^t when coloring all t-hyperedges of the complete graph on $R_t(k;r)$ vertices. Since our arbitrary coloring assigns colors to all subsets of size t, we see that all t-hyperedges are the same color, say color i_t. Since t-hyperedges are subsets of t vertices, taking A to be the set of vertices of this K_k^t finishes the proof since $A \subseteq C$ inherits the subset color property of C and $|A| = k$. □

Before getting to the proof of Theorem 2.55, the following lemma allows us some flexibility. Recall Definition 2.9 of $FS(s_1, s_2, \ldots, s_k)$, which is the set of finite sums of a sequence (as opposed to a set).

Lemma 3.22. *Assume that for every $k \in \mathbb{Z}^+$, every r-coloring of $\mathbb{Z}^+$ admits a sequence of k (not necessarily distinct) integers $s_1, s_2, \ldots, s_k$ such that $FS(s_1, s_2, \ldots, s_k)$ is monochromatic. Then, for any $\ell \in \mathbb{Z}^+$, every r-coloring of $\mathbb{Z}^+$ admits a set S of ℓ elements such that $FS(S)$ is monochromatic.*

Proof. Consider the hypothesis of the lemma. By the Compactness Principle, there exists an integer $n(k;r)$ such that every r-coloring of $[1, n(k;r)]$ admits a monochromatic $FS(s_1, s_2, \ldots, s_k)$. To show the conclusion of the lemma, we will show that every r-coloring of $[1, n(\ell^3;r)]$ admits a monochromatic $FS(S)$ with $|S| = \ell$.

Let $s_1, s_2, \ldots, s_{\ell^3}$ be a sequence of ℓ^3 (not necessarily distinct) integers that satisfies the hypothesis. We may assume that there are not ℓ distinct integers in this sequence; otherwise, we are done. Hence, each s_i takes on one of at most $\ell - 1$ values. By the pigeonhole principle, one of these values must have at least ℓ^2 members of the sequence. Hence, let v be a value taken on by ℓ^2 terms of the sequence: renumbering if necessary, let these be $s_1, s_2, \ldots, s_{\ell^2}$. Then

$$\{v, 2v, \ldots, \ell^2 v\} = FS(s_1, s_2, \ldots, s_{\ell^2})$$

is monochromatic. Let $S = \{iv : i = 1, 2, \ldots, \ell\}$. Then

$$FS(S) = \left\{v, 2v, \ldots, \binom{\ell+1}{2} v\right\} \subseteq \{v, 2v, \ldots, \ell^2 v\},$$

so $FS(S)$ is monochromatic with $|S| = \ell$, and we are done. □

With this lemma, we now need not worry about whether or not the elements that exist are distinct, provided the hypothesis of Lemma 3.22 is satisfied.

Proof of Theorem 2.55. By applying Lemma 3.22 it suffices to show that for any $k \in \mathbb{Z}^+$ there exists an integer $f(k;r)$ such that every r-coloring of $[1, f(k;r)]$ admits k (not necessarily distinct) integers $s_1, s_2, \ldots, s_k$ such that $FS(s_1, s_2, \ldots, s_k)$ is monochromatic. We prove this via induction on k where the base case $k = 2$ is Schur's Theorem. Hence, we assume $f(k-1;r)$ exists for all r.

Let $N(k,t;r)$ be as defined in Lemma 3.21 and let

$$n = N(4f(k-1;r), 4f(k-1;r);r).$$

We will show that $f(k;r) \le n$. To start, define the following function on the subsets of $[1,n]$:

$$\text{for } B = \{b_1 < b_2 < \cdots < b_m\} \subseteq [1,n], \text{ let } g(B) = (-1)^m \sum_{i=1}^{m} (-1)^i b_i.$$

Note that $1 \le g(B) \le n$ for any $B \ne \emptyset$.

Let $\chi : [1,n] \to \{1,2,\ldots,r\}$ be an arbitrary coloring. Using χ we define a coloring $\gamma : \wp([1,n]) \setminus \emptyset \to \{1,2,\ldots,r\}$ by

$$\gamma(B) = \chi(g(B)).$$

Applying Lemma 3.21, under γ there exists $A \subseteq [1,n]$ of size $4f(k-1;r)$ and a sequence of $4f(k-1;r)$ colors $i_1, i_2, \ldots, i_{4f(k-1;r)}$ such that all subsets of A of size j have color i_j, for $1 \le j \le 4f(k-1;r)$. This assures us that the r-coloring τ of $4[1, f(k-1;r)]$ defined by $\tau(c) = \gamma(C) = i_c$, where C is any subset of A with c elements, is well-defined. (That is not a typo; τ colors multiples of 4.)

Since sumsets are invariant under dilation of an interval, we may identify τ as an r-coloring of $[1, f(k-1;r)]$ so that we may use the inductive assumption. Hence, we have $t_1 \le t_2 \le \cdots \le t_{k-1}$ all being integers belonging to $4[1, f(k-1;r)]$ such that $FS(t_1, t_2, \ldots, t_{k-1})$ is monochromatic under τ. Note that this implies that

$$\sum_{i=1}^{k-1} t_i \le 4f(k-1;r),$$

which is important since the elements of $FS(t_1, t_2, \ldots, t_{k-1})$ will be playing the role of indices of terms in A, which is a set of size $4f(k-1;r)$.

Define the partial sums $p_0 = 0$ and

$$p_j = \sum_{i=1}^{j} t_i$$

for $j = 1, 2, \ldots, k-1$. Denote the set A by $\{a_1 < a_2 < \cdots < a_{4f(k-1;r)}\}$. Consider the sets

$$B_j = \{a_{p_{j-1}+1}, a_{p_{j-1}+2}, \ldots, a_{p_j}\}$$

for $j = 1, 2, \ldots, k-1$ and let

$$B_k = \left\{a_{\frac{p_1}{2}}, a_{\frac{p_1}{2}+1}, \ldots, a_{p_1-1}\right\} \bigsqcup \left\{a_{p_1+2}, a_{p_1+3}, \ldots, a_{\frac{3p_1}{2}+1}\right\}.$$

Note that $|B_j| = t_j$ for $1 \leq j \leq k-1$ and $|B_k| = t_1$.

We now go back to our function g and consider the sequence:

$$s_j = g(B_j) = \sum_{i=p_{j-1}+1}^{p_j} (-1)^i a_i$$

for $j = 1, 2, \ldots, k-1$ and

$$s_k = g(B_k) = \sum_{i=\frac{p_1}{2}}^{p_1-1} (-1)^{i-1} a_i + \sum_{i=p_1+2}^{\frac{3p_1}{2}+1} (-1)^{i-1} a_i.$$

We will show that $s_1, s_2, \ldots, s_k$ satisfies the theorem, thereby finishing the induction (and, hence, proof).

We start by showing that $s_1, s_2, \ldots, s_k$ is monochromatic under our original coloring χ by tracing back through the various colorings. For $1 \leq j \leq k-1$ we have

$$\chi(s_j) = \chi(g(B_j)) = \gamma(B_j) = \tau(|B_j|) = \tau(t_j).$$

Since $t_1, t_2, \ldots, t_{k-1}$ is a monochromatic sequence under τ we have that $s_1, s_2, \ldots, s_{k-1}$ is a monochromatic sequence under χ. Similarly, since $\chi(s_k) = \tau(t_1)$ we now see that $s_1, s_2, \ldots, s_k$ are colored identically under χ.

Continue by considering

$$s(I) = \sum_{i \in I} s_i$$

with $\emptyset \neq I \subseteq \{1, 2, \ldots, k\}$. First, consider any I with $k \notin I$. In this case, all the associated sets B_j are disjoint so that

$$\chi(s(I)) = \chi\left(g\left(\bigsqcup_{i \in I\setminus\{k\}} B_i\right)\right) = \gamma\left(\bigsqcup_{i \in I\setminus\{k\}} B_i\right) = \tau\left(\left|\bigsqcup_{i \in I\setminus\{k\}} B_i\right|\right) = \tau\left(\sum_{i \in I\setminus\{k\}} t_i\right),$$

which, by the inductive assumption, is the same color as the monochromatic sumset $FS(t_1, t_2, \ldots, t_{k-1})$. Hence, $s(I)$ has the same color as $\{s_i\}_{i=1}^k$.

It remains to show that $s(I) = \sum_{i \in I} s_i$ with $\emptyset \neq I \subseteq \{1, 2, \ldots, k\}$ and $k \in I$ is also the same color as $\{s_i\}_{i=1}^k$. If neither 1 nor 2 are elements of I then we again have disjoint sets and can conclude, as above (noting that $|B_k| = t_1$), that $s(I)$ has the same color.

Next, assume $1 \in I$ but $2 \notin I$. Let

$$\widehat{B} = \left\{a_1, a_2, \ldots, a_{\frac{p_1}{2}-1}\right\} \bigsqcup \{a_{p_1}\} \bigsqcup \left\{a_{p_1+2}, a_{p_1+3}, \ldots, a_{\frac{3p_1}{2}+1}\right\}.$$

Then we have

$$s_1 + s_k = g(\widehat{B})$$

and $\widehat{B} \cap B_j = \emptyset$ for $3 \leq j \leq k-1$. Hence, we now have disjoint sets as before. Since

$$|\widehat{B}| = p_1 = t_1$$

we have

$$\sum_{i \in I} s_i = (s_1 + s_k) + \sum_{i \in I \setminus \{1,k\}} s_i$$

so that

$$\chi(s(I)) = g\left(\widehat{B} \sqcup \bigsqcup_{i \in I \setminus \{1,k\}} B_i\right) = \tau\left(t_1 + \sum_{i \in I \setminus \{1,k\}} t_i\right).$$

Hence, in this case $s(I)$ is the same color as $\{s_i\}_{i=1}^k$.

The case when $2 \in I$ but $1 \notin I$ is almost identical, so we move on to $1, 2 \in I$. Defining

$$\widetilde{B} = \left\{a_1, a_2, \ldots, a_{\frac{p_1}{2}-1}\right\} \bigsqcup \{a_{p_1}, a_{p_1+1}\} \bigsqcup \left\{a_{\frac{3p_1}{2}+2}, a_{\frac{3p_1}{2}+3}, \ldots, a_{p_2}\right\},$$

we have

$$s_1 + s_2 + s_k = g(\widetilde{B}).$$

Furthermore, $\widetilde{B}$ has size $p_2 - p_1 = t_2$ and is disjoint from all B_j with $3 \leq j \leq k-1$. Thus,

$$\sum_{i \in I} s_i = (s_1 + s_2 + s_k) + \sum_{i \in I \setminus \{1,2,k\}} s_i$$

so that

$$\chi(s(I)) = \tau\left(t_2 + \sum_{i \in I \setminus \{1,2,k\}} t_i\right)$$

and is the same color as the sequence $\{s_i\}_{i=1}^k$.

Hence, we have shown that all elements of $FS(s_1, s_2, \ldots, s_k)$ have the same color under χ, as needed. □

3.3.3 Symmetric Hypergraph Theorem

Here we present a useful tool for determining bounds on Ramsey-type functions. We will be using graph statistics that appear in Section 3.2.1, specifically the chromatic number and the independence number as they apply to hypergraphs.

Definition 3.23 (Chromatic number of a hypergraph)**.** Let G be a hypergraph. The *chromatic number of* G, denoted $\chi(G)$, is the minimal number of colors r needed so that we can color the vertices of G in such a way that every hyperedge has at least two vertices of different colors.

Definition 3.24 (Independence number of a hypergraph, Independent set)**.** Let $G = (V, E)$ be a hypergraph. The *independence number of* G, denoted $\alpha(G)$, is the maximal size of a subset of vertices $W \subseteq V$ such that no hyperedge in E consists of only vertices from W (i.e., G contains a subhypergraph on $\alpha(G)$ vertices with no hyperedge). We will refer to W as an *independent set.*

In order to state and apply this subsection's titular theorem, we have need of the following definition.

Definition 3.25 (Automorphism group, Symmetric hypergraph)**.** Let $G = (V, E)$ be a hypergraph (or graph). Let $\mathrm{Sym}(V)$ be the symmetric group of V (i.e., permutations of the elements in V). The *automorphism group of* G, denoted $\mathrm{Aut}(G)$, is a subgroup of $\mathrm{Sym}(V)$ with the property that for all $\pi \in \mathrm{Aut}(G)$ we have $\{i_1, i_2, \ldots, i_\ell\} \in E$ if and only if $\{\pi(i_1), \pi(i_2), \ldots, \pi(i_\ell)\} \in E$. We say that G is a *symmetric hypergraph* if $\mathrm{Aut}(G)$ acts transitively on V; that is, if for any $u, v \in V$ there exists $\pi \in \mathrm{Aut}(G)$ such that $\pi(u) = v$.

Remark. The symmetry of a symmetric hypergraph is more technically about vertex-symmetry; however, the Ramsey theory literature does not make this distinction.

As a trivial example of a symmetric hypergraph, we see that K_n^ℓ is symmetric for all values of ℓ and n.

Theorem 3.26 (Symmetric Hypergraph Theorem)**.** *Let* $G = (V, E)$ *with* $|V| = n$ *be a symmetric hypergraph. Then*

$$\frac{n}{\alpha(G)} \leq \chi(G) < 1 + \frac{n}{\alpha(G)} \ln n,$$

where G *need not be a symmetric hypergraph to apply the lower bound.*

Proof. The lower bound holds easily: we can color the vertices with $\chi(G)$ colors in a way that every hyperedge has vertices of different colors. We deduce that some color, say c, is used at least $\frac{n}{\chi(G)}$ times. Those vertices of color c must be a set with no hyperedge defined on them. Hence,

$$\frac{n}{\chi(G)} \leq \alpha(G).$$

We now prove the upper bound. We will show that

$$n\left(1-\frac{\alpha(G)}{n}\right)^{\chi(G)-1} \geq 1,$$

from which a little algebra, and the use of $\ln(1-x) < -x$, yields the stated upper bound. In order to deduce this, we need a preliminary bound (given in Inequality (3.2), below).

Let $H = \mathrm{Aut}(G)$ so that all elements in H preserve hyperedges. Let $I \subseteq V$ be a maximal independent set (so that $|I| = \alpha(G)$).

We will show that for any $U \subseteq V$, there exists $\pi \in H$ such that

$$|\pi I \cap U| \geq \frac{\alpha(G)}{n}|U|. \tag{3.2}$$

To prove this, let $i \in I$ and $u \in U$ and consider all $\sigma \in H$ such that $\sigma(i) = u$. Let H_i be the stabilizer of i, i.e.,

$$H_i = \{\tau \in H : \tau(i) = i\}.$$

Since i is fixed by every element in H_i, we see that every element in the coset σH_i takes i to u. Furthermore, if σ_1 and σ_2 both send i to u, then we have $\sigma_2 = \sigma_1\tau$ for some $\tau \in H_i$ by considering $\tau(j) = \sigma_1^{-1}\sigma_2(j)$ for all $j \in V$.

From the above argument, to determine the number of elements of H that send i to u we need to determine how many elements are in the coset σH_i. To do so, we apply the Orbit-Stabilizer Theorem (Theorem 1.12). Since H acts transitively on V we know that the orbit of any $i \in V$ is all of V. Hence, from the Orbit-Stabilizer Theorem, we find that (the index) $[H : H_i] = |V| = n$. From Lagrange's Theorem (see Section 1.3.4) we have $[H : H_i] = \frac{|H|}{|H_i|} = n$ so that

$$|H_i| = \frac{|H|}{n}.$$

Since all (left) cosets of H_i have the same number of elements, we see that there are $\frac{|H|}{n}$ elements $\sigma \in H$ such that $\sigma(i) = u$ for each fixed pair of i and u.

Hence, by summing over possible $u \in U$ we have

$$\sum_{\sigma \in H} |\{\sigma(i)\} \cap U| = \frac{|H||U|}{n}.$$

Summing over all $i \in I$ we have

$$\sum_{\sigma \in H} |\sigma I \cap U| = \frac{|H||I||U|}{n} = \frac{\alpha(G)|U|}{n}|H|.$$

From this we can deduce that for some specific $\pi \in H$ we must have

$$|\pi I \cap U| \geq \frac{\alpha(G)}{n}|U|;$$

otherwise, the sum over all elements of H could not equal $\frac{\alpha(G)|U|}{n}|H|$.

We now apply Inequality (3.2) iteratively. Let $U_0 = V$ and take $\pi_0 \in H$ so that Inequality (3.2) is satisfied. For $i = 1, 2, \ldots$, let

$$U_i = U_{i-1} \setminus (\pi_{i-1} I \cap U_{i-1}),$$

where $\pi_{i-1} \in H$ is chosen so that Inequality (3.2) is satisfied with $U = U_{i-1}$. From this, it follows that

$$|U_i| \leq \left(1 - \frac{\alpha(G)}{n}\right) |U_{i-1}|,$$

so that

$$|U_m| \leq \left(1 - \frac{\alpha(G)}{n}\right)^m |V| = n\left(1 - \frac{\alpha(G)}{n}\right)^m \tag{3.3}$$

for any $m \in \mathbb{Z}^+$. Let r be the minimal positive integer satisfying $|U_r| = 0$ but $|U_{r-1}| \geq 1$, which must occur since

$$\left(1 - \frac{\alpha(G)}{n}\right)^j < \frac{1}{n}$$

for j sufficiently large. Then we can write

$$V = \bigsqcup_{i=0}^{r-1} \pi_i I,$$

which defines an r-coloring of V. Since each $\pi_i \in \text{Aut}(G)$ preserves hyperedges and I is an independent set, we see that no $\pi_i I$ contains a hyperedge. Hence, we must be using at least $\chi(G)$ colors since our coloring of the vertices has the property that each hyperedge has at least two vertices of different colors. Thus, $r \geq \chi(G)$. Since $|U_{r-1}| \geq 1$ we have $|U_{\chi(G)-1}| \geq 1$ so that, by Inequality (3.3), we have

$$1 \leq n\left(1 - \frac{\alpha(G)}{n}\right)^{\chi(G)-1},$$

which is what we needed to show. □

Remark. Only the upper bound is referred to as the Symmetric Hypergraph Theorem; however, we have chosen to place the lower bound in for ease of application, and to show the range of possible $\chi(G)$ values for symmetric hypergraphs.

In [84] the authors state that the Symmetric Hypergraph Theorem is a "folklore" theorem and present what is believed to be the first published proof of it. The proof above follows their proof (and attempts to locate its origin have been unsuccessful).

To finish this section, we will apply the Symmetric Hypergraph Theorem to some Ramsey-type numbers. The bounds obtained are not the best-known as these examples are intended for illustrative purposes only.

Example 3.27. Let $n = R(3;r) - 1$ so that there exists an r-coloring of the edges of K_n that avoids monochromatic triangles. Construct the 3-uniform hypergraph $G = (V, E)$ where $V = \{\{i,j\} : 1 \leq i < j \leq n\}$ and

$$E = \{\{\{i,j\} \cup \{i,k\} \cup \{j,k\}\} : 1 \leq i < j < k \leq n\}.$$

In other words, the vertices of G are all possible pairs of positive integers that are at most n and the hyperedge set consists of those triples of pairs that only use three different integers. Note that our vertex set is the set of edges of K_n and our set of hyperedges is the set of triangles in K_n.

Let's assume that G is a symmetric hypergraph and determine what bound we can obtain (no need to do more work than needed). Under this (perhaps faulty) assumption, by the Symmetric Hypergraph Theorem, we first note that $\chi(G) = r$ by the definition of n. We know that $|V| = \binom{n}{2}$ so it remains to find a bound for $\alpha(G)$. To this end, consider $W \subseteq V$ defined by

$$W = \left\{\{i,j\} : 1 \leq i \leq \frac{n}{2} < j \leq n\right\}.$$

This must be an independent set since if $\{i_1, j_1\}$ and $\{i_1, j_2\}$ are both in W then $\{j_1, j_2\}$ is not in W (and $\{i_1, i_2\} \notin W$ for $\{i_1, j_1\}$ and $\{i_2, j_1\}$). Since $|W| \approx \left(\frac{n}{2}\right)^2 = \frac{n^2}{4}$ we have

$$\alpha(G) \geq \frac{n^2}{4}.$$

Applying Theorem 3.26 we obtain

$$\chi(G) = r < 1 + \frac{\binom{n}{2}}{\frac{n^2}{4}} \ln\left(\binom{n}{2}\right) \approx 1 + 4\ln n - 2\ln 2.$$

From this we conclude that

$$n > \sqrt{2}\, e^{\frac{r-1}{4}}$$

so that

$$R(3;r) > \sqrt{2}\, e^{\frac{r-1}{4}} - 1.$$

Although this is a nontrivial bound, it is far from the best-known lower bound for $R(3;r)$ and is included only as an illustration for applying (partially) the Symmetric Hypergraph Theorem. The current best asymptotic bound is

$$R(3;r) \geq (3.1996)^r$$

(see Rowley [176]; also Chung [43]). Hence, determining whether or not G is symmetric would not be worthwhile since even with $|I| \approx \binom{n}{2}$ we still only obtain $R(3;r) > \sqrt{2}e^{\frac{r-1}{2}}$.

We now present an application to the other main Ramsey-type numbers, the van der Waerden numbers, where we also give details on proving a hypergraph is symmetric.

Example 3.28. Let $n = w(k;r) - 1$ and assume we have an r-coloring of $[1, n]$ with no monochromatic k-term arithmetic progression. We start with the obvious hypergraph: let $G = (V, E)$ be the hypergraph on $[1, n]$ where every k-term arithmetic progression in $[1, n]$ is a hyperedge and there are no other hyperedges. The obvious members of $\text{Sym}(V)$ that preserve arithmetic progressions are the shift permutations: $\pi_c : i \mapsto i + c$ for $c \in [1, n - 1]$. Unfortunately, these are not in $\text{Aut}(G)$ since, for example, if we consider the k largest elements of $[1, n]$ (which would be a hyperedge) and apply π_1 we obtain $\{1\} \cup [n - k + 2, n]$, which is not a hyperedge of G.

The shift permutations are clearly the permutations that preserve arithmetic progressions, so we need to determine how to handle the "wrap-around" arithmetic progressions.

A first thought could be to construct a superhypergraph H for which these wrap-around arithmetic progressions are edges. This would mean that our hyperedges are k-term arithmetic progressions in the additive group $\mathbb{Z}_n$. Clearly G is a subgraph of H. But we again run into an issue. Letting $\alpha(G)$ and $\alpha(H)$ be the corresponding independence numbers, we have $\alpha(G) \geq \alpha(H)$ as H has more edges on the same vertex set. If we apply the Symmetric Hypergraph Theorem to H, we have

$$\chi(H) \leq 1 + \frac{n}{\alpha(H)} \ln n.$$

To deduce a bound for G, since $\chi(G) \leq \chi(H)$ we have

$$r = \chi(G) \leq 1 + \frac{n}{\alpha(H)} \ln n.$$

Since $\alpha(G) \geq \alpha(H)$ we can go no further. In order to conclude that

$$1 + \frac{n}{\alpha(H)} \ln n \leq 1 + \frac{n}{\alpha(G)} \ln n$$

we need our superhypergraph H to satisfy $\alpha(H) \geq \alpha(G)$.

So, our goal now is to find a superhypergraph with the property that these wrap-around arithmetic progressions are not contained in $[1, n]$. The remaining option is to increase the vertex set. So, consider H on the vertex set $[1, 2n]$ instead. The idea is to have two "copies" of G in H with hyperedges added so that H is symmetric. Every arithmetic progression $a, a + d, \ldots, a + (k - 1)d$ in $[1, n]$ satisfies $(k - 1)d < n$. So, we define $H = ([1, 2n], E')$ where

$$E' = \left\{ \{a, a + d, \ldots, a + (k - 1)d\} \subseteq \mathbb{Z}_{2n} : d < \frac{n}{k - 1} \right\}.$$

We will now show that H is a symmetric hypergraph. Let $\pi_c \in \text{Sym}([1, 2n])$ be a shift permutation. Then $e \in E'$ if and only if $\pi_c(e) \in E'$ is clear so that $\pi_c \in \text{Aut}(H)$. Furthermore, any "wrap-around" arithmetic progression is not entirely contained in $[1, n]$. Hence, we can now conclude that $\alpha(G) \leq \alpha(H)$ as an independent set of vertices in G is also independent in H. Clearly the set of shift permutations is transitive on $[1, 2n]$. Thus, H is a symmetric hypergraph. We can now apply the upper bound in Theorem 3.26 to H along with $\chi(G) \leq \chi(H)$ and $\alpha(G) \leq \alpha(H)$ to conclude that

$$r = \chi(G) \leq \chi(H) \leq 1 + \frac{n}{\alpha(H)} \ln n \leq 1 + \frac{n}{\alpha(G)} \ln n.$$

By definition, we have $\alpha(G) = r_k(n)$ (see the notation on page 61). Hence,

$$r \leq \frac{n}{r_k(n)} \ln n(1 + o(1)). \tag{3.4}$$

The current best lower bound for $r_k(n)$ is due to O'Bryant [155]:

$$r_k(n) \geq \frac{cn(\log n)^{\frac{1}{2\ell}}}{2^{\ell \cdot 2^{\frac{\ell-1}{2}}}(\log n)^{\frac{1}{\ell}}},$$

where $\ell = \lceil k \log k \rceil$, for some constant $c > 0$.

Using this bound in Inequality (3.4) and recalling that $n = w(k; r) - 1$ yields

$$w(k; r) > r^{cd_k(\log r)^{\ell-1}},$$

where $d_k = \left(\ell \cdot 2^{\frac{\ell-1}{2}}\right)^{-\ell}$, for some $c > 0$, which is not as strong (for fixed r) as the cr^{k-1} bound on $w(k; r)$ given in Theorem 2.4.

We can also use $\chi(G) \geq \frac{n}{\alpha(G)}$ from Theorem 3.26 to provide an upper bound. We still have $\alpha(G) = r_k(n)$. The best upper bound for $r_k(n)$, for general k, comes from Gowers [82] and this is precisely how we get the upper bound in Theorem 2.4. We do have improvement over the general $r_k(n)$ when we specify to $k = 3$. Inequality (2.28) gives

$$r_3(n) \ll \frac{n}{(\log n)^{1+c}}$$

for some fixed constant $c > 0$. Hence, we have

$$r \gg (\log n)^{1+c},$$

from which we deduce that

$$w(3; r) < 2^{r^{1-\epsilon}}(1 + o(1))$$

for some $\epsilon > 0$, from which we deduce that $w(3; r)$ is sub-exponential.

3.4 Infinite Graphs

The prototypical infinite graph is the complete graph on $\mathbb{Z}$ (or $\mathbb{Z}^+$ if you prefer), where every pair of integers is an edge. Abstracting to hypergraphs, we use the following notation.

Notation. Let $\ell \in \mathbb{Z}^+$. The complete ℓ-uniform hypergraph on vertex set V is denoted K_V^ℓ. The hyperedge set consists of all sets of ℓ integers. For $\ell = 2$ and $V = \mathbb{Z}$ we rely on the notation $K_\mathbb{Z}$.

We start by considering r-colorings of the edges of $K_\mathbb{Z}$. Of particular note here is that the number of colors we are using is finite. Under this condition, Ramsey's Theorem informs us that any such coloring admits a monochromatic K_k, where k can be arbitrarily large. This does not imply that we have a monochromatic complete graph on an infinite vertex set. For example, consider the 2-coloring of $K_{\mathbb{Z}^+}$ where all edges between vertices $2^i, 2^i+1, \dots, 2^{i+1}-1$ for all $i \in \mathbb{Z}^+$ are one color, say red, and all other edges are the other color. We have arbitrarily large red complete graphs, but clearly no infinite red complete graph.

Ramsey [166] is also responsible for the next theorem.

Theorem 3.29 (Infinite Ramsey Theorem). *Let $\ell, r \in \mathbb{Z}^+$. Every r-coloring of the ℓ-hyperedges of $K_\mathbb{Z}^\ell$ admits a vertex set V with $|V| = \infty$ such that all ℓ-hyperedges of K_V^ℓ are the same color.*

Proof. We will prove this via induction on ℓ. Technically, we may start with $\ell = 1$ (which holds trivially as some color must be used an infinite number of times); however, the $\ell = 2$ case will be more instructive, so we start with it.

Let $\chi : E(K_\mathbb{Z}) \to \{c_1, c_2, \dots, c_r\}$ be an arbitrary coloring. Isolate a vertex v_1. For some color, say c_{i_1}, vertex v_1 is connected to an infinite number of vertices V_2 by color c_{i_1}. Pick $v_2 \in V_2$ and consider the edges from v_2 to $V_2 \setminus \{v_2\}$. For some color, say c_{i_2}, vertex v_2 is connected to an infinite number of vertices $V_3 \subseteq V_2 \setminus \{v_2\}$ by color c_{i_2}. Continue in this fashion to obtain the sequence $v_1, v_2, v_3, \dots$ with $v_i \in V_i$ and $V_i \subseteq V_{i-1} \setminus \{v_{i-1}\}$ for $i \geq 2$. By construction, for any $j < k$ we have v_j connected to v_k by color c_{i_j}.

Among the colors $c_{i_1}, c_{i_2}, \dots$ some color must occur an infinite number of times, say $c_{i_{k_1}} = c_{i_{k_2}} = c_{i_{k_3}} = \cdots$. Let

$$V = \bigsqcup_{j=1}^{\infty} \{v_{i_{k_j}}\}$$

and note that all edges between pairs of vertices in V have the same color. This completes the $\ell = 2$ case.

We now assume the result holds for $\ell - 1$ and will show that it holds for ℓ. Let $\chi : E(K_\mathbb{Z}^\ell) \to \{c_1, c_2, \dots, c_r\}$ be an arbitrary ℓ-hyperedge-coloring. Let $w_0 \in \mathbb{Z}$ be arbitrary and define $\widehat{\chi}_0$ to be the r-coloring of the $(\ell-1)$-hyperedges e of $K_{\mathbb{Z}\setminus\{w_0\}}^{\ell-1}$ given by

$$\widehat{\chi}_0(e) = \chi(e \cup \{w_0\}).$$

By the inductive assumption, $\widehat{\chi}_0$ admits an infinite complete subhypergraph $G_1 = (W_1, E_1)$ all of whose $(\ell-1)$-hyperedges are the same color, say c_0. Hence, for all $e \in E_1$, we see that $e \cup \{w_0\}$ is an ℓ-hyperedge of $K_\mathbb{Z}^\ell$ of color c_0 under χ.

Pick $w_1 \in W_1$ and define $\widehat{\chi}_1$ to be the r-coloring of the $(\ell-1)$-hyperedges e of

$$K^{\ell-1}_{W_1\setminus\{w_0,w_1\}}$$

given by $\widehat{\chi}_1(e) = \chi(e \cup \{w_1\})$. Using the inductive assumption, we again find an infinite complete subhypergraph $G_2 = (W_2, E_2)$ all of whose $(\ell-1)$-hyperedges are the same color, say c_1. Hence, for all $e \in E_2$, we see that $e \cup \{w_1\}$ is an ℓ-hyperedge of $K^{\ell}_{\mathbb{Z}}$ of color c_1 under χ.

By construction, $W_2 \subseteq W_1$ so that $G_2 \subseteq G_1$. By defining $\widehat{\chi}_2, \widehat{\chi}_3, \dots$ in a similar fashion, we have the existence of infinite vertex sets $W_1, W_2, W_3, \dots$ with associated monochromatic $(\ell-1)$-hyperedge sets $E_1, E_2, E_3, \dots$ with the property that $W_{i+1} \subseteq W_i$ and $G_{i+1} \subseteq G_i$ for all $i \in \mathbb{Z}^+$. For every $(\ell-1)$-hyperedge $e \in E_{i+1}$ we have that $\chi(e \cup \{w_i\}) = c_i$ for some $w_i \in W_i$. Among the c_i some color must occur an infinite number of times. Thus, we can conclude that $c_{i_1} = c_{i_2} = c_{i_3} = \cdots$. Consider $W = \{w_{i_1}, w_{i_2}, \dots\}$. By construction, all ℓ-hyperedges of W have color c_{i_1} under χ, thereby proving the theorem. □

As we can see, the above proof hinges on there being a finite number of colors. Does a similar result hold if we allow an infinite number of colors? Clearly it cannot be the same result if we use each of an infinite number of colors only once. Perhaps by avoiding monochromatic infinite subhypergraphs, we cannot avoid infinite subhypergraphs where every hyperedge has a distinct color. This avenue is explored next.

3.4.1 Canonical Ramsey Theorem

Our first-blush guess at the end of the last section is not far from the truth. By going through the proof of Theorem 3.29 with the caveat that we may have an infinite number of colors, we will be able to deduce what is known as the Canonical Ramsey Theorem (Theorem 3.30, below). We will not offer a formal proof; rather we will point out the implications of having an infinite number of colors as we analyze the proof of Theorem 3.29 to show that the situations in the Canonical Ramsey Theorem are natural.

We will only consider colorings of $K_{\mathbb{Z}}$. The abstraction to $K^{\ell}_{\mathbb{Z}}$ results in the consideration of 2^{ℓ} possible situations; see [130] for details.

In the $\ell = 2$ case of the proof of Theorem 3.29, if there exists some color, say c_{i_j}, such that vertex v_j is connected to an infinite number of vertices V_{j+1} by color c_{i_j} at each stage, then we can conclude that either there is a monochromatic K_V or that $\chi(v_i v_j) = \chi(v_k v_\ell)$ if and only if $i = k$ depending on whether or not we have, among the colors $c_{i_1}, c_{i_2}, \dots$, some color that occurs an infinite number of times.

In the situation when no color connects v_j to an infinite number of vertices, we have an infinite number of colors each occurring a finite number of times. Furthermore, this must hold for every vertex in V_j (otherwise, we could just choose a different vertex and continue as before). Hence, we can assume that for all $w \in V_j$, the vertex is connected to the other vertices in V_j by infinitely many distinct colors, each color occurring only a finite number of times.

We will now make use of Theorem 3.29 by defining a 4-coloring γ: $E(K^3_{V_j}) \to \{0,1,2,3\}$ of the 3-hyperedges of V_j. Denoting V_j by $\{v_1, v_2, \dots\}$ and considering an arbitrary hyperedge $e = \{v_i, v_j, v_k\}$, $i < j < k$, we define

$$\begin{aligned}
\gamma(e) &= 0 \quad \text{if } \chi(v_iv_k) = \chi(v_jv_k) \neq \chi(v_iv_j);\\
\gamma(e) &= 1 \quad \text{if } \chi(v_iv_j), \chi(v_iv_k), \chi(v_jv_k) \text{ are distinct;}\\
\gamma(e) &= 2 \quad \text{if } \chi(v_iv_j) = \chi(v_iv_k) = \chi(v_jv_k);\\
\gamma(e) &= 3 \quad \text{if } \chi(v_iv_j) = \chi(v_iv_k) \neq \chi(v_jv_k).
\end{aligned}$$

We can conclude that there exists an infinite subset $W \subseteq V_j$ such that all 3-hyperedges of W are monochromatic under γ. We know this color is neither 2 nor 3 as we have ruled them out at this stage. This leaves only colors 0 and 1 as possibilities, which is reflected in the statement of the final theorem of this section.

Theorem 3.30 (Canonical Ramsey Theorem). *Color by χ the edges of $K_{\mathbb{Z}}$ with an arbitrary (perhaps infinite) number of colors. Then there exists an infinite set of vertices $V = \{v_1, v_2, \dots\}$ such that one of the following holds:*

(i) K_V is monochromatic;

(ii) All $\chi(v_iv_j)$, $1 \leq i < j$, are distinct;

(iii) For all $i < j$ and $k < \ell$ we have $\chi(v_iv_j) = \chi(v_kv_\ell)$ if and only if $i = k$;

(iv) For all $i < j$ and $k < \ell$ we have $\chi(v_iv_j) = \chi(v_kv_\ell)$ if and only if $j = \ell$.

Theorem 3.30 can also be described as follows: For some $I \subseteq \wp(\{1,2\})$ we have $\chi(v_{i_1}v_{i_2}) = \chi(v_{j_1}v_{j_2})$ if and only if $v_{i_k} = v_{j_k}$ for all $k \in I$. In this reframing, the extension of Theorem 3.30 to $K^\ell_{\mathbb{Z}}$ would be stated as: For some $I \subseteq \wp(\{1,2,\dots,\ell\})$ we have $\chi(v_{i_1}v_{i_2}\cdots v_{i_\ell}) = \chi(v_{j_1}v_{j_2}\cdots v_{j_\ell})$ if and only if $v_{i_k} = v_{j_k}$ for all $k \in I$. (In these restatements we assume $i_1 < i_2 < \cdots$ and $j_1 < j_2 < \cdots$.)

3.5 Comparing Ramsey and van der Waerden Results

In the last section we showed that Ramsey's Theorem, which proves arbitrarily large monochromatic complete subgraphs, can be extended to prove infinitely large monochromatic complete subgraphs (provided we use a finite number of colors). As we have noted before, although van der Waerden's Theorem proves arbitrarily long monochromatic arithmetic progressions, we cannot guarantee an infinitely long monochromatic arithmetic progression (consider the 2-coloring of $\mathbb{Z}^+$ given by coloring all integers in

$$\bigsqcup_{i=0}^{\infty}[2^{2i}, 2^{2i+1}-1]$$

one color and its complement the other color).

We have also seen that van der Waerden's Theorem has a density analogue: if $\limsup_{n\to\infty} \frac{|A\cap[1,n]|}{n} > 0$ then A contains arbitrarily long arithmetic progressions. What would an analogous density statement for Ramsey's Theorem be? Since we are coloring edges, as a start the density should be relative to the $\binom{n}{2}$ edges in K_n. Perhaps if we consider a graph with edge set A satisfying $|A| > \epsilon n^2$ for some $\epsilon > 0$ we can guarantee arbitrarily large complete graphs with edges from A. Unfortunately, this fails.

Consider V_i with $|V_i| = \frac{n}{k-1}$ for $i = 1, 2, \ldots, k-1$ and let $V = \bigsqcup_{i=1}^{k-1} V_i$. Define the complete $(k-1)$-partite graph $T = T(n, k-1)$ on V with an edge between any $v \in V_i$ and any $w \in V_j$ provided $i \neq j$ (and no other edge). Then T does not contain K_k as a subgraph since any such subgraph must have two vertices from the same V_i (meaning there is no edge between these two vertices). The number of edges in T is

$$\binom{k-1}{2}\left(\frac{n}{k-1}\right)^2 \approx \frac{k-2}{k-1}\cdot\frac{n^2}{2}$$

so that the density of edges in T relative to the $\binom{n}{2}$ possible edges of K_n is $\frac{k-2}{k-1} = 1 - \frac{1}{k-1}$. Hence, for all $\epsilon > 0$, the graph $T(n, 1+\epsilon^{-1})$ has an edge density of $1-\epsilon$ without containing arbitrarily large complete subgraphs (as $n \to \infty$).

Thus, we can have an edge density arbitrarily close to 1 without a guarantee of arbitrarily large complete subgraphs. The graphs used to show this are called Turán graphs.

Definition 3.31 (Turán graph). Let $n, \ell \in \mathbb{Z}^+$. The *Turán graph* $T(n, \ell)$ partitions n vertices into ℓ equal (as best as possible) sets with edges between any two vertices in different sets and no other edge.

The graph $T(n, k)$, although belonging to the class of complete k-partite graphs, is given this name as it arises from the proof of a theorem bearing Turán's name. We present here a graph-theoretic proof; a different proof of the inequality given in the theorem can be found in Section 6.2.

Theorem 3.32 (Turán's Theorem). *Let $G = (V, E)$ be a graph on n vertices with no K_k subgraph. Then*

$$|E| \leq \frac{k-2}{2(k-1)}n^2.$$

This bound is sharp and is attained by $T(n, k-1)$, the unique such graph.

Proof. Let G be a graph on n vertices with no K_k subgraph with the maximal number of edges. We will first show that G must be a complete $(k-1)$-partite graph.

By edge maximality, there exists a K_{k-1} in G. Let $W = \{w_1, w_2, \ldots, w_{k-1}\}$ be the vertices of this complete graph. Since G does not contain a K_k, every vertex in $V \setminus W$ has at most $k-2$ edges to vertices in W. We conclude, by maximality, that every such vertex has exactly $k-2$ edges to W.

For every $v \in V \setminus W$, let $f(v) = w \in W$, where there is no edge between v and w. Note that because v has $k-2$ edges to W, such a w is unique.

Let

$$V_i = \{v \in V \setminus W : f(v) = w_i\} \cup \{w_i\}.$$

Notice that we now have the partition

$$V = \bigsqcup_{i=1}^{k-1} V_i.$$

Consider $u, v \in V_i$ for any given i. By construction, each of u and v are connected to every vertex in $W \setminus \{w_i\}$. Hence, we cannot have an edge between u and v, for otherwise the graph on $\{u, v\} \cup (W \setminus \{w_i\})$ would be a K_k subgraph of G. We conclude that no two vertices in the same V_i share an edge.

At this point, we have shown that G is $(k-1)$-partite. To see that it is a complete $(k-1)$-partite graph, notice that because G has the maximal number of edges, each $v \in V_i$ must be connected to every $u \notin V_i$ since a K_k subgraph requires two vertices in the same V_j, which is not possible.

We now claim that we must have $|V_i| = |V_j| + \delta$ with $\delta \in \{0, \pm 1\}$ for all $1 \leq i, j \leq k-1$. To see this, assume otherwise, i.e., that $|V_i| \geq |V_j| + 2$ for some $i \neq j$. Consider the graph G' created by taking $x \in V_i$ (which is contained in $n - |V_i|$ edges in G), deleting all edges connected to x, moving x to V_j, and creating all edges between x and every vertex not in V_j. These added edges number

$$n - (|V_j| + 1) \geq n - |V_i| + 1,$$

which is more than the $n - |V_i|$ we removed. Hence, $E(G') > E(G)$, which is not possible since G contains the maximal number of edges.

At this stage we have shown that $G = T(n, k-1)$. The proof is completed by noting that a complete $(k-1)$-partite graph with vertices' partition sizes of $\left\lfloor \frac{n}{k-1} \right\rfloor + \epsilon$, $\epsilon \in \{0, 1\}$ has at most

$$\binom{k-1}{2} \left(\frac{n}{k-1} \right)^2 = \left(1 - \frac{1}{k-1} \right) \frac{n^2}{2}$$

edges. □

So far we have seen that Ramsey's Theorem and van der Waerden's Theorem are fundamentally different with respect to whether or not infinite monochromatic structures exist, and whether or not a density analogue holds. We have also seen that Ramsey's Theorem has a hypergraph analogue. In this direction, what would an appropriate analogue for van der Waerden's Theorem be? One obvious direction is that, since a hypergraph has edges of higher dimension, we should consider multi-dimensional arithmetic progressions. If we consider the arithmetic progressions $a_1, a_1+d, a_1+2d, \ldots, a_1+kd$ and $a_2, a_2+d, a_2+2d, \ldots, a_1+kd$ to be the projections onto the x-axis and y-axis, respectively, from a 2-dimensional arithmetic progression, then our 2-dimensional arithmetic progression would be a subset of $\{(a_1, a_2) + d(i, j) : 0 \le i, j \le k\}$. The observant reader will notice that the common gap d is the same for both projected arithmetic progressions. This is intentional as otherwise we would have two different arithmetic progressions and the name 2-dimensional arithmetic progression would be a misnomer.

Definition 3.33. Let $k, m \in \mathbb{Z}^+$. For $d \in \mathbb{Z}^+$ and $(a_1, a_2, \ldots, a_m) \in (\mathbb{Z}^+)^m$, the set

$$\{(a_1, a_2, \ldots, a_m) + d(i_1, i_2, \ldots, i_m) : 0 \le i_1, i_2, \ldots, i_m \le k-1\} \subseteq (\mathbb{Z}^+)^m$$

is called an *m-dimensional arithmetic progression of dimension length k*.

The fact that we can guarantee the existence of monochromatic multi-dimensional arithmetic progressions under all finite colorings follows from the Gallai-Witt Theorem, below, which relies on the Hales-Jewett Theorem. The proof is similar to how we deduce van der Waerden's Theorem as a corollary of the Hales-Jewett Theorem (see Section 2.3).

Theorem 3.34 (Gallai-Witt Theorem). *Let $m, r \in \mathbb{Z}^+$ and let $S \subseteq \mathbb{Z}^m$ be a finite set. There exists $n = n(S, m; r)$ such that for any r-coloring of the points in $[-n, n]^m \subseteq \mathbb{Z}^m$, there exists $\mathbf{a} \in \mathbb{Z}^m$ and $d \in \mathbb{Z}^+$ such that $\mathbf{a} + dS$, i.e.,*

$$\{\mathbf{a} + d\mathbf{s} : \mathbf{s} \in S\},$$

is monochromatic.

Proof. Consider the Hales-Jewett number $h = HJ(|S|; r)$ so that every r-coloring of the length h words over S (our alphabet) admits a monochromatic variable word, i.e., for some $I \neq \emptyset$, the variable word

$$\vec{w}(\mathbf{x}) = (\mathbf{w_1}, \mathbf{w_2}, \ldots, \mathbf{w_h}) \text{ with } \mathbf{w_i} = \mathbf{x} \text{ for } i \in I \text{ and } \mathbf{w_i} \in S \text{ otherwise,}$$

has the property that the r-coloring is constant on $\{\vec{w}(\mathbf{s}) : \mathbf{s} \in S\}$.

For every $\mathbf{s} = (s_1, s_2, \ldots, s_m) \in S$, let $j(\mathbf{s}) = \max_{1 \le i \le m} |s_i|$ and define $n = h \max_{\mathbf{s} \in S} j(\mathbf{s})$.

Let χ be an arbitrary r-coloring of $[-n, n]^m$ and define the r-coloring γ of

S^h (i.e., the words of length h over S) by

$$\gamma(\vec{s}) = \chi\left(\sum_{\mathbf{s_i}\in\vec{s}} \mathbf{s_i}\right).$$

In other words, we take a length h word, which consists of h vectors of m integers each, and perform the usual addition of vectors to obtain a vector of size m with all entries in $[-n, n]$.

By the Hales-Jewett Theorem, there exists a monochromatic (under γ) variable word $\vec{u}(\mathbf{x})$. Let d equal the number of occurrences of the variable $\mathbf{x}$ in $\vec{u}(\mathbf{x}) = (\mathbf{u_1}, \mathbf{u_2}, \ldots, \mathbf{u_h})$ and note that $d \in \mathbb{Z}^+$. Then, under χ, we see that

$$\left\{\sum_{\mathbf{u_i}\neq\mathbf{x}} \mathbf{u_i} + d\,\mathbf{s} : \mathbf{s} \in S\right\}$$

is monochromatic. Letting $\mathbf{a} = \sum_{\mathbf{u_i}\neq\mathbf{x}} \mathbf{u_i}$ completes the proof. □

As an immediate corollary, by taking

$$S = \{(i_1, i_2, \ldots, i_m) : 0 \leq i_1, i_2, \ldots, i_m \leq k-1\},$$

we have our analogue of the Hypergraph Ramsey Theorem in the arithmetic progression setting.

Corollary 3.35 (Multi-dimensional van der Waerden Theorem). *Let $k, m, r \in \mathbb{Z}^+$. There exists $n = n(k, m; r)$ such that for any r-coloring of $[1, n]^m$ there exists $a_1, a_2, \ldots, a_m, d \in \mathbb{Z}^+$ such that the m-dimensional arithmetic progression of dimension length k*

$$\{(a_1, a_2, \ldots, a_m) + d(i_1, i_2, \ldots, i_m) : 0 \leq i_1, i_2, \ldots, i_m \leq k-1\}$$

is monochromatic.

The last area of exploration when comparing Ramsey's Theorem and van der Waerden's Theorem concerns the Canonical Ramsey Theorem. We know that we cannot guarantee a monochromatic infinite length arithmetic progression even when using only two colors, so any type of canonical theorem concerning arithmetic progressions needs to apply to finite-length arithmetic progressions. So the question becomes, what can we say if we allow infinitely many colors?

Theorem 3.36 (Canonical van der Waerden Theorem). *Let $k \in \mathbb{Z}^+$. Color $\mathbb{Z}^+$ by χ with an arbitrary (perhaps infinite) number of colors. Then there exist $a, d \in \mathbb{Z}^+$ such that one of the following holds:*

(i) all $\chi(a + id)$, $0 \leq i \leq k-1$, are identical;

(ii) all $\chi(a+id)$, $0 \leq i \leq k-1$, are distinct.

Proof. By van der Waerden's Theorem, if we restrict to colorings with a finite number of colors, then (i) always holds. Hence, we need only consider colorings with an infinite number of colors.

We will apply Corollary 3.35. In order to do so, we need to construct an appropriate finite coloring. To this end, for every k-term arithmetic progression $b, b+c, \ldots, b+(k-1)c$, let t_i, $1 \leq i \leq k$, be the number of terms of color $\chi(b+(i-1)c)$. Then $(t_1, t_2, \ldots, t_k) \in [1,k]^k$. Lastly, define a k^k-coloring of $(\mathbb{Z}^+)^2$ by

$$\gamma((b,c)) = (t_1, t_2, \ldots, t_k).$$

This is valid since (b,c) uniquely determines the k-term arithmetic progression $b, b+c, \ldots, b+(k-1)c$.

Let $n = n(2k^2+1, 2; k^k)$ be from Corollary 3.35 so that

$$A = \{(a_1,a_2) \pm d(i,j) : 0 \leq i,j \leq k^2\} \subseteq [1,n]^2 \subseteq (\mathbb{Z}^+)^2$$

is monochromatic under γ (where we have centered the 2-dimensional arithmetic progression at (a_1,a_2) for ease of presentation). If A has color $(1,1,\ldots,1)$, then, in particular, $\gamma(a_1,a_2) = (1,1,\ldots,1)$. This means that all colors of $a_1, a_1+a_2, a_1+2a_2, \ldots, a_1+(k-1)a_2$ are distinct, so that (ii) holds. Hence, we assume that A has a different color. In particular, some entry in $\gamma(a_1,a_2)$ is at least 2. So, in particular, at least two terms of $a_1, a_1+a_2, a_1+2a_2, \ldots, a_1+(k-1)a_2$ have the same color. Assume $\chi(a_1+ra_2) = \chi(a_1+sa_2)$ are both, say, red, with $0 \leq r < s < k$.

As A is monochromatic with respect to γ, considering $\gamma((a_1+di, a_2+dj))$ for fixed i and $|j| \leq k$, we can conclude that $\chi(a_1+di+r(a_2+dj)) = \chi(a_1+di+s(a_2+dj))$.

Specializing to $i = -rj$, we have $\chi(a_1+ra_2) = \chi(a_1+sa_2+(s-r)dj)$. Since $\chi(a_1+ra_2)$ is red, we can conclude that $\chi(a_1+sa_2+(s-r)dj)$ is also red for every $j \in \{0,1,\ldots,k\}$. Thus, $(a_1+sa_2), (a_1+sa_2)+(s-r)d, (a_1+sa_2)+2(s-r)d, \ldots, (a_1+sa_2)+(k-1)(s-r)d$ is a red k-term arithmetic progression, which is situation (i) of the theorem. □

3.6 Exercises

3.1 Prove that $R(k,k)$ exists using only the existence of $R(k-1,k-1)$. In other words, show that the diagonal Ramsey numbers exists without the existence of off-diagonal Ramsey numbers.

Hint:

Consider edge-wise 2-colorings of K_n with $n = 2(k-2)R(k-1;2)+2$.

3.2 Denote by $\mathcal{E}(n)$ the equation $\sum_{i=1}^{n-1} x_i = x_n$. Let $k, \ell \in \mathbb{Z}^+$ with $k, \ell \geq 3$. Show that there exists an integer $m = m(k, \ell)$ such that every 2-coloring of $[1, m]$ admits a monochromatic solution to $\mathcal{E}(k)$ of the first color or a monochromatic solution to $\mathcal{E}(\ell)$ of the second color.

3.3 Show that every 2-coloring of the edges of K_6 admits two monochromatic triangles.

3.4 Show that every 2-coloring of the edges of K_n admits at least $\frac{n^3}{24}(1 + o(1))$ monochromatic triangles. (This is due to Goodman [81].)

3.5 The following result is due to Lefmann [132]. Let $k, r, s \in \mathbb{Z}^+$. Show that

$$R(k; r + s) > (R(k; r) - 1)(R(k; s) - 1).$$

Hint:

Let $n = R(k; r) - 1$ and $m = R(k; s) - 1$. Let the vertices of K_n be $\{(0, j) : j = 1, 2, \ldots, n\}$ and let the vertices of K_m be $\{(i, 0) : i = 1, 2, \ldots, m\}$. Consider K_{n+m} with vertices $\{(i, j) : 1 \leq i \leq n, 1 \leq j \leq m\}$.

3.6 Let $r \in \mathbb{Z}^+$. Prove that for every r-coloring of $\mathbb{Z}^+$ there exists a monochromatic infinite sequence $s_1 < s_2 < \cdots$ such that all $s_i + s_{i+1}$ have the same color as the sequence. Use only results from this chapter and do not appeal to Hindman's Theorem (Theorem 2.56).

3.7 Show that

$$R(k_1, k_2, \ldots, k_r) \leq \binom{\sum_{i=1}^{r} k_i}{k_1, k_2, \ldots, k_r}.$$

Deduce that

$$R(k; r) \leq \frac{r^{rk+\frac{1}{2}}}{(2\pi k)^{\frac{r-1}{2}}}(1 + o(1)).$$

3.8 One of the two Ramsey numbers known for more than 2 colors is $R(3, 3, 3) = 17$. Use cosets of cubic residues in the finite field of 2^4 elements to prove that $R(3, 3, 3) > 16$. Deduce that $R(3, 3, 3) = 17$ via the pigeonhole principle.

3.9 The only hypergraph Ramsey number known is $R_3(4, 4) = 17$. The determination of this requires a computer; however, $R_3(4, 4) \leq 19$ does not require a computer. Show that $R_3(4, 4) \leq 19$.

3.10 Determine the chromatic number of the Peterson graph (see Figure 3.2) as well as the minimal number of colors needed to color the edges of this graph so that no vertex has two edges of the same color incident to it (this minimum is called the *chromatic index* or *edge chromatic number*). Also, determine the independence number of this graph.

3.11 Let G be a finite graph such that $\chi(G) = k$. Prove that G is a k-partite graph.

3.12 Show that every 2-coloring of the *vertices* of K_n admits at least $\frac{n^3}{24}(1+o(1))$ triangles with all vertices the same color. Extend this to more than 2 colors.

3.13 Show that $R(C_4, C_4) = 6$ and $R(C_3, C_4) = 7$.

3.14 Show that $R(K_{1,3}, K_{1,3}) = 6$ and $R(K_{1,4}, K_{1,4}) = 7$.

3.15 Consider the set of k-term arithmetic progressions in $[1, n]$. Call this set V and create the graph $G = (V, E)$, where an edge exists between $u, v \in V$ if and only if $u \cap v \neq \emptyset$ (i.e., the arithmetic progressions intersect). Let d be the maximal degree over all vertices. Show that $d < 2kn$.

3.16 Let $r \in \mathbb{Z}^+$ and color every subset of $[1, n]$ with one of r colors, i.e., r-color the elements of $\wp([1, n])$. Prove or disprove: for $n = n(k; r)$ sufficiently large, for any such coloring there exists a monochromatic sequence of subsets $S_1, S_2, \ldots, S_k$ in $\wp([1, n])$ with the property that $|S_{i+1} \setminus S_i| = 1$ for all $i \in [1, k-1]$.

3.17 Prove the following without appealing to the Multi-dimensional van der Waerden Theorem (Corollary 3.35): Let $k, \ell, r \in \mathbb{Z}^+$. For every r-coloring of $(\mathbb{Z}^+)^2$ there exists a k-term arithmetic progression A and an ℓ-term arithmetic progression B such that $A \times B$ is monochromatic. Note that this is weaker than the Multi-dimensional van der Waerden Theorem as the arithmetic progressions' common differences need not be equal.

3.18 Consider the grid $\{x = i : i \in \mathbb{Z}^+\} \times \{y = i : i \in \mathbb{Z}^+\}$. In each unit square defined by this grid place an integer. Prove that for any $n \in \mathbb{Z}^+$, there exists a square S consisting of a union of unit squares such that the sum of the integers in S is divisible by n. This result is a particular case of a result from [16].

Hint:

Color the unit square with lower left vertex (i, j) by the sum of values in all unit squares (k, ℓ) with $k \leq i$ and $\ell \leq j$ modulo n.

4

Euclidean Ramsey Theory

She parted with convention, now she's living in the 4th dimension without me.

–Mark Mothersbaugh & Gerald Casale

This chapter continues the study of graphs and colorings from a slightly different angle. In the last chapter, we considered colorings of vertices, edges, and hyperedges. In this chapter, we will by coloring Euclidean spaces, e.g., the plane, and trying to deduce monochromatic structures that must exist within the space, or what we can say about the colorings when structures are imposed on them.

As a simple introduction, color all points in $\mathbb{R}^2$ with one of two colors, say, red or blue. For any $d \in \mathbb{R}^+$ there must exist two points at a distance d apart of the same color. To see this, just consider an equilateral triangle of side length d (clearly two vertices must be the same color).

If we seek a similar argument with 3 colors, we can consider a regular tetrahedron in $\mathbb{R}^3$ with side length d. However, if we restrict to 3-colorings of $\mathbb{R}^2$, this simple problem becomes a bit trickier and is left to the reader as Exercise 4.3.

4.1 Polygons

Relying on monochromatic points at a distance d in 2-colorings of $\mathbb{R}^2$, we can show that there exists an isosceles triangle with all vertices the same color. We may assume that v and x are both red at a distance d, where x is on the circle centered at v of radius d. All other points on this circle must be blue or we are done (except, perhaps, the point on the circle colinear with x and v, but we need not consider this). Let $y \neq x$ be nearby x on this circle (so that y is blue) such that the distance between x and y is less than d. Consider the circle centered at y of radius d and let a and b be the points of intersection of the two circles. Since a and b are on the first circle, they cannot be red. Hence, both a and b are blue and Δaby is an isosceles triangle with all vertices blue. Figure 4.1 gives a "proof without words" version of this argument.

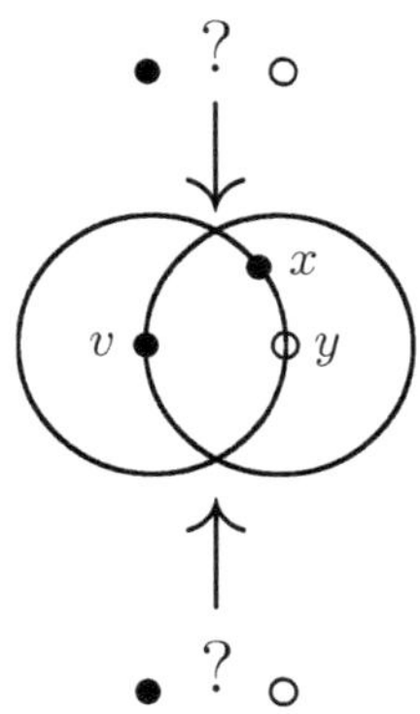

FIGURE 4.1
Proof without words of isosceles triangles in 2-colorings of $\mathbb{R}^2$ with all vertices the same color

Remark. Throughout this chapter, when considering polygons with all vertices the same color, we will refer to the polygon as monochromatic instead of the more cumbersome "with all vertices of the same color." Also, since we are not coloring edges of graphs in this chapter, when we refer to a polygon we are referring to the vertices of the polygon.

Showing that a monochromatic isosceles triangle exists in any 2-coloring of $\mathbb{R}^2$ is quite easy. Is the same true for more than two colors? Can we guarantee other monochromatic polygons exist? We will investigate some such questions in this section.

We will rely on the following definition (based on one found in [84]).

Definition 4.1. Let $n, r \in \mathbb{Z}^+$. Let $P \subseteq \mathbb{R}^n$ be a set of polygons. We let $E(P, n; r)$ be the statement

> $E(P, n; r)$: *every r-coloring of $\mathbb{R}^n$ admits a monochromatic polygon q with $q \cong p$ for some $p \in P$.*

If P is given as a sequence of r elements, i.e., $E((p_1, p_2, \ldots, p_r), n; r)$, the statement is interpreted as

> $E((p_1, p_2, \ldots, p_r), n; r)$: *every r-coloring of $\mathbb{R}^n$ admits a monochromatic polygon q of color i, for some $i \in \{1, 2, \ldots, r\}$, with $q \cong p_i$.*

We start by showing how the Gallai-Witt Theorem (Theorem 3.34) addresses certain types of sets P in Definition 4.1. Consider any polygon described by its vertices, say $v_1, v_2, \ldots, v_t \in \mathbb{Z}^n$. Taking $S = \{v_1, v_2, \ldots, v_t\}$

in Theorem 3.34 we see that any r-coloring of $\mathbb{Z}^n$, and hence $\mathbb{R}^n$, admits a monochromatic set of the form $\mathbf{a} + dS$. This means that we have a dilated copy of our polygon that is monochromatic under any finite coloring of $\mathbb{R}^n$. Thus, if we let P in Definition 4.1 be a set of all polygons similar to a given polygon, the Gallai-Witt Theorem guarantees that $E(P, n; r)$ is true.

Having this similarity result, an obvious next step is to let P be a single polygon. We start by letting T be the following equilateral triangle of side length one:

$$T = \left\{ (0,0), (1,0), \left(\frac{1}{2}, \frac{\sqrt{3}}{2} \right) \right\}.$$

To see that $E(T, 2; 2)$ is false, color $\mathbb{R}^2$ by horizontal half-open strips of height $\frac{\sqrt{3}}{2}$, closed at the lower edge and open at the upper edge. This argument can easily be modified to allow T to be any equilateral triangle. We leave the details as part of Exercise 4.2.

It may be surprising that, although $E(T, 2; 2)$ is false when T is any equilateral triangle, if S is any right triangle, then $E(S, 2; 2)$ is true. This result was shown by Shader [188].

Theorem 4.2. *Let S be a right triangle. Then every 2-coloring of the points of $\mathbb{R}^2$ admits a monochromatic T with $T \cong S$.*

We will rely on Lemmas 4.3 and 4.4, below, found in [65], which we will not prove.

Lemma 4.3. *Let $a, b, c \in \mathbb{R}^+$ and let T be a triangle with side lengths a, b, and c. Let χ be a 2-coloring of $\mathbb{R}^2$. Then χ admits a monochromatic copy of T if and only if χ admits a monochromatic copy of an equilateral triangle of side length a, b, or c.*

Lemma 4.4. *Let $a, b, c \in \mathbb{R}^+$. Let $k \in \mathbb{Z}^+$ be odd and denote by S_k the right triangle with side lengths ka and kb and hypotenuse length kc. Let χ be a 2-coloring of $\mathbb{R}^2$. If χ admits a monochromatic copy of S_k, then χ admits a monochromatic copy of S_1.*

Proof of Theorem 4.2. Let $a, b \in \mathbb{R}^+$ be arbitrary side lengths of right triangle S. We will prove the following claim:

`Claim.` Every 2-coloring of $\mathbb{R}^2$ admits a monochromatic equilateral triangle of side length ka for some $k \in \{1, 3, 5, 7, 13, 15\}$.

Upon showing this, by Lemma 4.3 we have a monochromatic right triangle with side lengths ka and kb. Then, by Lemma 4.4, we have a monochromatic right triangle with side lengths a and b, i.e., a monochromatic copy of S, which is to be shown.

The key to proving the claim is to note that in a $3-5-7$ triangle, the angle between the sides of lengths 3 and 5 is $120°$ and that in a $7-13-15$ triangle, the angle between the sides of lengths 7 and 15 is $60°$.

Using the colors red and blue, let R be the set of red points and B be the set of blue points. We denote by E_s an equilateral triangle with side length s.

First, we may assume that not all equilateral triangles of side length 8 are monochromatic. To see this, assume otherwise and let u be a red point. Then all points on the circle of radius 8 centered at u must be red. Let v and w be points on this circle one unit apart. As shown in Figure 4.2, let x be outside of the circle so that $\Delta vwx \cong E_1$. Consider an E_8 with one vertex x and one vertex y that intersects the circle (let z be the other vertex). Since y is red and we assuming that all E_8 are monochromatic, we have that x must be red. But then Δvwx is a red E_1. Hence, if all equilateral triangles of side length 8 are monochromatic, then the claim is true.

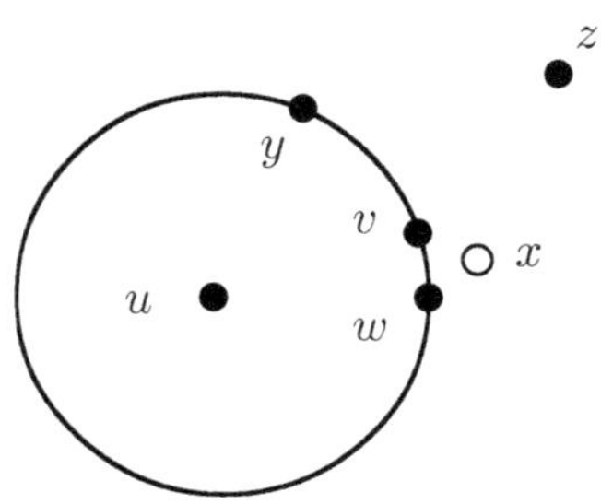

FIGURE 4.2
Not all E_8 are monochromatic

We will prove the claim by contradiction so we may also assume, by Lemmas 4.3 and 4.4, that no monochromatic $3-5-7$ or $7-13-15$ triangles exist. We will use a skewed coordinate system to describe points (skewed at 60°; see Figure 4.3).

We may assume, without loss of generality, that $(0,0) \in R$ and $(8,0),(0,8) \in B$ since not all E_8 are monochromatic. Furthermore, since $(0,3),(3,0)$, and $(0,8)$ form a $3-5-7$ triangle, one of $(0,3)$ and $(3,0)$ must be red. We may also assume, without loss of generality, that $(0,3) \in R$. This gives us that $(3,0) \in B$ (otherwise $\{(0,0),(0,3),(3,0)\}$ is a red E_3). By considering the $3-5-7$ triangle $\{(0,0),(0,3),(5,3)\}$ we see that $(5,3) \in B$. The equilateral triangle $\{(5,0),(8,0),(5,3)\}$ gives us $(5,0) \in R$, which, in turn, gives $(0,5) \in B$. To avoid $\{(0,5),(0,8),(3,5)\}$ and $\{(0,5),(0,8),(-3,8)\}$ being monochromatic, we have $(3,5),(-3,8) \in R$. As a final initial deduction we show that $(1,7) \in R$: Since $\{(-5,5),(0,5),(0,8)\}$ is a $3-5-7$ triangle, we have $(-5,5) \in R$. Coupling $(-5,5)$ with $(0,0)$ gives us that $(-5,0) \in B$. As $\{(-5,0),(8,0),(1,7)\}$ is a $7-13-15$ triangle, we obtain $(1,7) \in R$.

These initial deductions are exhibited in Figure 4.3 where the filled circles represent red points, while the hollowed circles represent blue points.

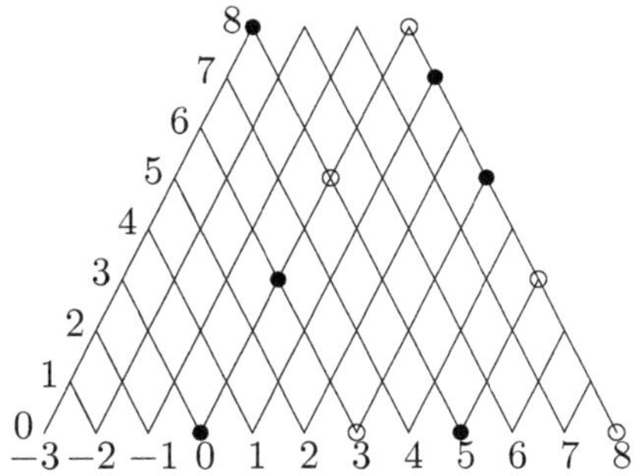

FIGURE 4.3
Initial deductions for the proof of Theorem 4.2

At this stage we need to consider cases. For our exposition, we will use a diagram similar to the one in Figure 4.3 to show the flow of implications. We will use subscripts on the colors and refer to them in a series of implications. Those with letter subscripts are from Figure 4.3.

`Case 1.` $(2,0) \in R$. Consider Figure 4.4 ($(2,0)$ is labeled as 0). By avoiding monochromatic equilateral triangles of odd side lengths as well as monochromatic $3-5-7$ triangles we have the following sequence of implications:

$0, \mathrm{c} \Rightarrow 1;$	$0, \mathrm{j} \Rightarrow 2;$	$1, \mathrm{f} \Rightarrow 3;$	$2, \mathrm{g} \Rightarrow 4;$	$3, \mathrm{i} \Rightarrow 5;$	$3, \mathrm{h} \Rightarrow 6;$
$5, 6 \Rightarrow 7;$	$2, 6 \Rightarrow 8;$	$4, 8 \Rightarrow 9;$	$5, 9 \Rightarrow 10;$	$10, \mathrm{e} \Rightarrow 11;$	$1, 11 \Rightarrow 12;$
$7, 12 \Rightarrow 13;$	$13, \mathrm{g} \Rightarrow 14;$	$14, \mathrm{e} \Rightarrow 15.$			

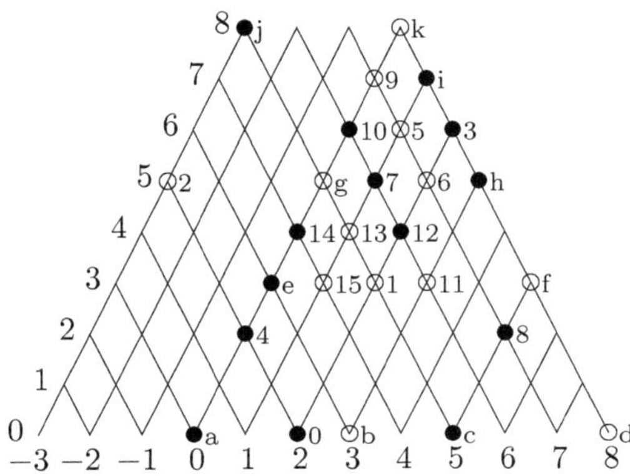

FIGURE 4.4
Case 1 of the proof of the claim

After this sequence of implications we deduce that $\{(1,3),(2,3),(1,4)\}$ is a blue E_1, finishing this case. ◇

`Case 2.` $(2,0) \in B$ and $(0,2) \in R$. Consider Figure 4.5 (in which $(2,0)$ is labeled as 0 and $(0,2)$ is labeled as 1). Consider the following sequence of implications:

1, e ⇒ 2, 3; 0, b ⇒ 4; 0, 3 ⇒ 5; 6 else $\{(1,3),(1,4),(2,3)\}$ is red;
2, 6 ⇒ 7; 5, 7 ⇒ 8; 4, 7 ⇒ 9; 9, b ⇒ 10; 8, 9 ⇒ 11.

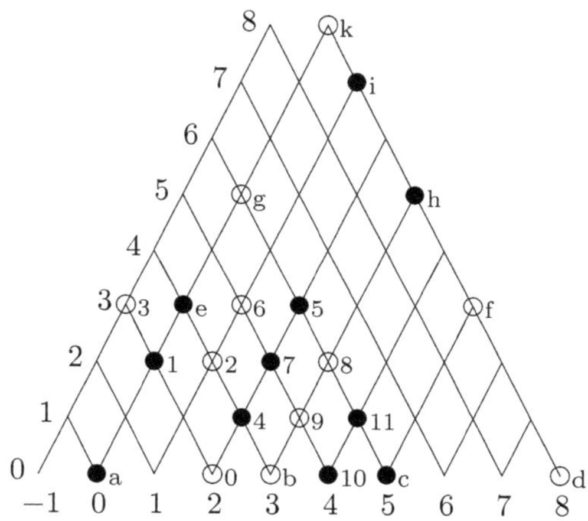

FIGURE 4.5
Case 2 of the proof of the claim

After this sequence of implications we deduce that $\{(4,0),(4,1),(5,0)\}$ is a red E_1, finishing this case. ◇

`Case 3.` $(0,2),(2,0) \in B$. Consider Figure 4.6 (in which $(2,0)$ is labeled as 0 and $(0,2)$ is labeled as 1). Consider the following sequence of implications:

a, e ⇒ 2; 0, b ⇒ 3; 1, g ⇒ 4; 3, 4, ⇒ 5; 4, h ⇒ 6;
5, 6 ⇒ 7; 7, a ⇒ 8; 0, 8 ⇒ 9, 10; 0, b ⇒ 11; 5, 6 ⇒ 12;
9, 12 ⇒ 13; 13, f ⇒ 14; 4, 14 ⇒ 15; 10, 11 ⇒ 16; 15, 16 ⇒ 17;
17, e ⇒ 18; 18, g ⇒ 19.

From this sequence of implications we deduce that $\{(1,3),(1,4),(2,3)\}$ is a red E_1, finishing this case. ◇

As the three cases exhaust all possibilities, the claim, and thereby the theorem, has been proved. □

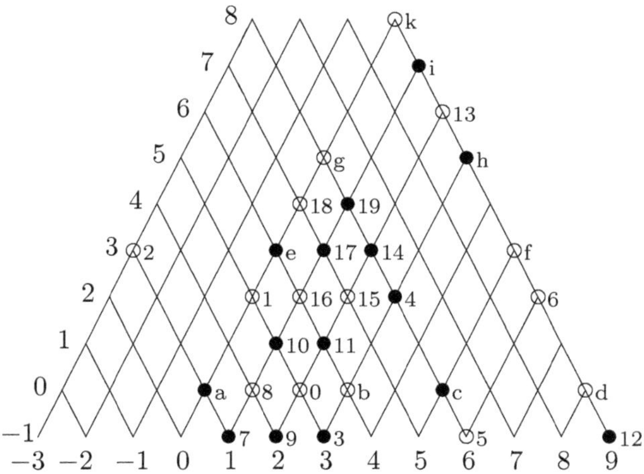

FIGURE 4.6
Case 3 of the proof of the claim

Unlike the tendency for a Ramsey statement to seemingly hold "more easily" in lower dimensions, in Euclidean Ramsey theory it is harder to have a monochromatic configuration in lower dimensions: $E(S, 2; r)$ being true for a given polygon S trivially implies that $E(S, 3; r)$ is true. For example, letting T be a right triangle with angles 30° and 60° and hypotenuse of length 1, $E(T, 2; 3)$ is currently unknown (Graham and Tressler conjecture in [85] that $E(S, 2; 3)$ is false for any triangle S), while Bóna [27] has shown that $E(T, 3; 3)$ is true (but $E(T, 3; 12)$ is false). A few years later, Bóna and Tóth [28] extended this to show that $E(T, 3; 3)$ is true and $E(T, 3; 21)$ is false for any right triangle T.

We know that if T is an equilateral triangle, then $E(T, 2; 2)$ is false. However, in [113] it is shown that every 2-coloring $R \sqcup B$ of $\mathbb{R}^2$ with one of R and B a closed subset (in the usual topology) admits any arbitrary triangle. For unrestricted colorings, in [64] it is shown that adding just one dimension affords us the ability to deduce the desired monochromatic structure.

Theorem 4.5. *Let T be an equilateral triangle. Then $E(T, 3; 2)$ is true.*

Proof. Without loss of generality, we may assume that T has side length 1. Using the colors red and blue, let P_1 and P_2 be (without loss of generality) red points at a distance 1 apart. Via translation and rotation we may assume that

$$P_1 = \left(-\frac{1}{2}, 0, 0\right) \quad \text{and} \quad P_2 = \left(\frac{1}{2}, 0, 0\right).$$

Consider the circle

$$C = \left\{(0, y, z) : y^2 + z^2 = \frac{3}{4}\right\}$$

and note that for any $P_3 \in C$ we have that $\Delta P_1P_2P_3 \cong T$. Hence, we may assume that all points in C are blue.

Let D be the set of midpoints of all pairs of points on C at a distance 1 apart. Hence, D is a circle of radius $\frac{\sqrt{2}}{2}$ concentric with C.

Let $Q_1, Q_2 \in C$ at a distance 1 apart with their midpoint on D so that we may assume there exists a circle E of radius $\frac{\sqrt{3}}{2}$ and perpendicular to $\overleftrightarrow{Q_1Q_2}$ of only red points, centered at a point in D. Letting F be the set of all such circles, we find that F is a self-intersecting torus (a spindle torus), all points of which are red. See Figure 4.7.

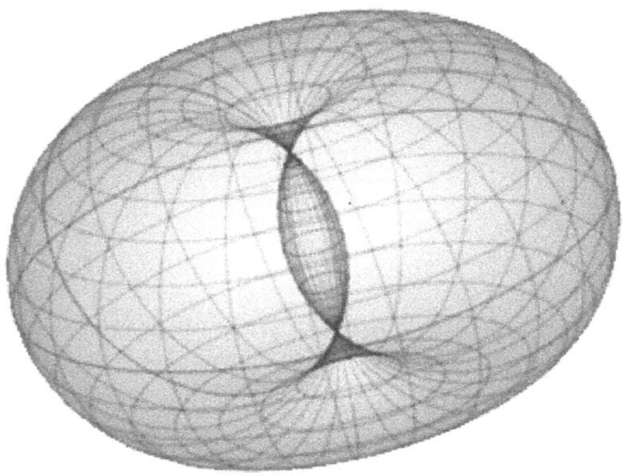

FIGURE 4.7
The spindle torus for proving monochromatic equilateral triangles in 2-colorings of $\mathbb{R}^3$

As evidenced from Figure 4.7, we have now deduced the existence of red circles (perpendicular to the line through the core of the torus) of all radii up to $\frac{\sqrt{2}+\sqrt{3}}{2}$. In particular, we have a circle of radius $\frac{\sqrt{3}}{3}$ consisting of only red points. On this circle, we have the vertices of an equilateral triangle of side length 1, thereby finishing the proof. □

Corollary 4.6. *For any triangle T, the statement $E(T, 3; 2)$ is true.*

Proof. Let $a \in \mathbb{R}^+$ be arbitrary and consider the equilateral triangle E_a of side length a. By Theorem 4.5 we have a monochromatic copy of E_a in any 2-coloring of $\mathbb{R}^3$. Consider the plane in which this monochromatic copy resides. Restricting to this plane we have (via rotation) a 2-coloring of $\mathbb{R}^2$ with a monochromatic copy of E_a. By Lemma 4.3, we are guaranteed any triangle with one side length a. Since a is arbitrary and this argument holds for any 2-coloring, the result follows. □

Moving from triangles to quadrilaterals, we know (as expected) much less. For example, if R is any given rectangle, $E(R, 2; 2)$ is unknown except in the case when it is a square: If S is a square, then $E(S, 2; 2)$ is false (see Exercise 4.2). With a simple constraint on the coloring, Juhász [115] has proved the following result. We omit the proof as it is fairly lengthy.

Theorem 4.7. *Let Q be any quadrilateral and let D be the unit line segment. Then $E((D,Q), 2; 2)$ is true.*

Note that we can rephrase Theorem 4.7 as: Every red-blue-coloring of $\mathbb{R}^2$ with no two points at unit distance apart being both red admits a blue quadrilateral. This result was recently strengthened in [114].

The best result for quadrilaterals without constraint on the coloring is due to Tóth [201] and is given next.

Theorem 4.8. *Let $n, m \in \mathbb{R}^+$ and denote by $R_{n,m}$ the rectangle of side lengths n and m. Then $E(R_{n,m}, 5; 2)$ is true.*

Proof. We may assume that $n \geq m$. Consider an arbitrary red-blue-coloring of $\mathbb{R}^5$. For our first step, we will show that we must have either a red $R_{n,m}$ or a blue (3-dimensional) regular tetrahedron of side length n. Let U be a blue point (if no such point, we are obviously done). Consider the 5-dimensional sphere of radius n centered at U. If this sphere is entirely red, then it contains a circle of diameter $\sqrt{n^2 + m^2}$. This circle contains the vertices of an $n \times m$ rectangle and, since it is entirely red, we are done. Hence, we let V be a blue point on this sphere. By changing coordinate systems, we may assume

$$U = \left(-\frac{n}{2}, 0, 0, 0, 0\right) \quad \text{and} \quad V = \left(\frac{n}{2}, 0, 0, 0, 0\right).$$

Consider the 4-dimensional sphere

$$S_1 = \left\{(0, x_2, x_3, x_4, x_5) \in \mathbb{R}^5 : x_2^2 + x_3^2 + x_4^2 + x_5^2 = \frac{3n^2}{4}\right\}.$$

For any point $P \in S_1$ we have $|UV| = |UP| = |VP| = n$. If S_1 is entirely red, then we again have a red circle of diameter $\sqrt{n^2 + m^2}$ in S_1 and are done. Hence, let

$$W = (0, w_2, w_3, w_4, w_5) \in S_1$$

be blue so that the triangle on vertices U, V, and W is a blue equilateral triangle with side length n.

Next, consider the 4-dimensional sphere of radius n centered at W:

$$S_2 = \left\{(0, x_2, x_3, x_4, x_5) \in \mathbb{R}^5 : (x_2 - w_2)^2 + \cdots + (x_5 - w_5)^2 = n^2\right\}.$$

Since S_1 contains all points in $\mathbb{R}^5$ at a distance n away from both U and V, while S_2 contains all points in $\mathbb{R}^5$ at a distance n away from W, we consider their intersection. Let

$$S_3 = S_1 \cap S_2.$$

If any point $X \in S_3$ is blue then $UVWX$ forms a blue 3-dimensional regular tetrahedron of side length n and we are done with our first step. Hence, we may assume S_3 is entirely red. A short calculation shows that S_3 is a 3-dimensional sphere with radius

$$\sqrt{\frac{2}{3}}n > \frac{n}{\sqrt{2}}.$$

Consequently, S_3 contains all circles of diameter at most $\sqrt{2}n$. Thus, since $n \geq m$, it contains a red circle of diameter $\sqrt{n^2+m^2}$, which admits a red $R_{n,m}$. This completes our first step.

At this point we have shown that every 2-coloring of $\mathbb{R}^5$ contains either a red $R_{n,m}$ or a blue 3-dimensional regular tetrahedron of side length n. If the former holds, we are done, so assume that our arbitrary coloring contains a blue 3-dimensional regular tetrahedron of side length n. Let $A, B, C,$ and D be the vertices of this tetrahedron and assume, without loss of generality, that

$$A = (0, 0, 0, 0, 0).$$

Let V be the 3-dimensional subspace of $\mathbb{R}^5$ spanned by $\overrightarrow{AB}, \overrightarrow{AC}$, and $\overrightarrow{AD}$ and let $V^{\perp}$ be the 2-dimensional subspace of $\mathbb{R}^5$ that is orthogonal to V.

Let $P_1, P_2 \in V^{\perp}$ such that ΔAP_1P_2 is an equilateral triangle of side length m. By orthogonality, $\overrightarrow{AP_1}, \overrightarrow{AP_2}$, and $\overrightarrow{P_1P_2}$ are all perpendicular to each edge of tetrahedron $ABCD$. Consider the tetrahedrons $P_iB_iC_iD_i$, $i = 1, 2$, where $B_i = B + P_i, C_i = C + P_i$, and $D_i = D + P_i$ (i.e., translate $ABCD$ by P_1 and also by P_2). Note that

$$|BB_i| = |CC_i| = |DD_i| = m.$$

If for either i we find that any two elements of $\{P_i, B_i, C_i, D_i\}$ are blue then we have a blue $R_{n\times m}$. For example, if P_1 and B_1 are blue, then ABB_1P_1 is a blue rectangle of the desired side lengths. Hence, we may assume that for each tetrahedron copy at most one vertex is blue. Thus, at least two of the pairs in $\{(P_1, P_2), (B_1, B_2), (C_1, C_2), (D_1, D_2)\}$ have both points red. Noting that P_1 and P_2 were chosen so that $|P_1P_2| = m$, we see that these red pairs form a red $R_{n,m}$, thereby finishing the proof. □

For many more results related to material in this section, see the wonderful "coloring" book by Soifer [191].

4.2 Chromatic Number of the Plane

In Theorem 4.7 we considered 2-colorings of $\mathbb{R}^2$ with no two points at unit distance apart both being a particular color. Clearly, as stated at the beginning of this chapter, we cannot avoid two points at a unit distance being the same color using only two colors. This leads to the question: If we use more colors, can we avoid two such points?

Consider the graph $G = (\mathbb{R}^2, E)$ where $\{x, y\} \in E$ if and only if $|x - y| = 1$. We call this the *unit distance graph.* Assuming that there exists a minimal number of colors we can use to avoid monochromatic points at a unit distance apart, we are attempting to determine the chromatic number of G (see Definition 3.12). The chromatic number of G, denoted $\chi(G)$, has come to be known as the *chromatic number of the plane.* Abusing notation, we use $\chi(\mathbb{R}^2)$ instead of $\chi(G)$. As with the lineage of many mathematical concepts, the history of this problem is a bit fuzzy, but it is typically referred to as the *Hadwiger-Nelson problem.* For a detailed history of the origins of this problem, the reader is pointed to Chapter 3 of [191].

As stated in the Introduction, we have attempted to stay away from Axiom of Choice issues. However, since we are dealing with $\mathbb{R}$, we need to consider it. The following theorem, due to de Bruijn and Erdős [54], allows us to consider finite graphs, remarkably allowing us to transition from arbitrarily large to infinite (something that the reader was cautioned against doing).

Theorem 4.9 (de Bruijn-Erdős Theorem). *Let G be a graph on infinitely many vertices. Let $\mathcal{H}$ be the family of finite subgraphs of G, i.e., of subgraphs with finite vertex sets. Assuming the Axiom of Choice, if $\chi(H) \leq r$ for all $H \in \mathcal{H}$ with equality attained, then $\chi(G) = r$.*

Because infinite Ramsey theory often requires separate tools from its finite counterpart, we will provide three different proofs. The one presented in this section relies only on Zorn's Lemma. In the next chapter, both a topological proof (as given by de Bruijn and Erdős) and an ultrafilter-based proof are provided, both of which also rely on the Axiom of Choice. The proof in this section is attributed to both Pósa (as noted by Lovász [137, p. 399], but without date) and Gabriel Dirac [59] (according to Komjáth [120], also without date), a student of Rado.

As we will be using Zorn's Lemma (which is equivalent to the Axiom of Choice), we restate it here for reference.

Lemma 4.10 (Zorn's Lemma). *If every chain in a partially ordered set S has an upper bound (respectively, lower bound) in S, then S contains a maximal (respectively, minimal) element.*

Proof of Theorem 4.9. Clearly $\chi(G) \geq r$, so we must show that $\chi(G) \leq r$. Let S be the set of all supergraphs of G on the same set of vertices, partially ordered by edge inclusion, with the property that every finite subgraph is r-colorable. Let $\mathcal{C}$ be a chain in this poset and consider

$$\mathcal{U} = \bigcup_{c \in \mathcal{C}} c.$$

We claim that $\mathcal{U}$ is an upper bound in S; in particular, we claim that $\mathcal{U}$ has the r-colorable property (since clearly it serves as an upper bound for $\mathcal{C}$). Suppose, for a contradiction, that $\mathcal{U}$ is not in S. Then $\mathcal{U}$ has a finite subgraph H that is not r-colorable. But then H must be a finite subgraph of some element of $\mathcal{C}$, a contradiction.

We can now apply Zorn's Lemma to obtain the existence of a maximal graph $G' \supseteq G$ so that the addition of any edge creates a finite subgraph that is not r-colorable. Let $G' = (V, E)$. Our goal is to show that there exists a partition $V = \bigsqcup_{i=1}^{d} V_i$ with $d \leq r$, such that $\{a, b\} \notin E$ if and only if $a, b \in V_i$ for some i (i.e., G' is d-partite; see Exercise 3.11). Under such a condition, we cannot have $d > r$ since then $K_{r+1} \subseteq G'$, which is clearly not r-colorable.

It suffices to show that $\sim$ given by $a \sim b$ if and only if $\{a, b\} \notin E$ is an equivalence relation on V so that it defines a partition of V and hence a vertex-valid coloring of the vertices. Reflexivity and symmetry of $\sim$ are obvious; only transitivity requires an argument. Let $a \sim b$ and $b \sim c$ and assume, for a contradiction, that $a \not\sim c$. By the maximality of G', if $\{a, b\}$ is inserted into E then there exists a finite subgraph S_1 that is no longer r-colorable. Similarly, if $\{b, c\}$ is appended to E we have a finite subgraph S_2 that is no longer r-colorable.

Consider $S_1 \cup S_2 \subseteq G'$. Clearly, a and c are both vertices of $S_1 \cup S_2$ and, by the assumption that $a \not\sim c$, we have that $\{a, c\}$ is an edge in $S_1 \cup S_2$. We will show that $S_1 \cup S_2$ is not r-colorable. Assume, for a contradiction, that $S_1 \cup S_2$ is r-colorable and let γ be a valid r-coloring of its vertices. By choice of S_1, we have $\gamma(a) = \gamma(b)$ since the inclusion of $\{a, b\}$ as an edge invalidates the r-colorability of S_1. Similarly, by choice of S_2 we have $\gamma(b) = \gamma(c)$. Hence, we see that $\gamma(a) = \gamma(c)$. Since $\{a, c\}$ is an edge in $S_1 \cup S_2$, we see that γ is not a vertex-valid r-coloring, a contradiction.

Thus far, under the assumption that $a \not\sim c$, we see that $S_1 \cup S_2$ is not r-colorable. However, $S_1 \cup S_2$ is a finite subgraph of G' and must, by choice of G', be r-colorable. Hence, the assumption that $a \not\sim c$ is false, proving that $\sim$ is an equivalence relation. This equivalence relation defines a coloring of the vertices of G' showing that G' is r-colorable. Since $G' \supseteq G$, we see that G is r-colorable so that $\chi(G) \leq r$, which was to be shown. □

With Theorem 4.9, we now have a tool to try to determine $\chi(\mathbb{R}^2)$. As a start, consider the following 7-vertex unit graph, known as the *Moser spindle* (see [149]), exhibited in Figure 4.8. The fact that this graph is not 3-colorable is left as Exercise 4.3.

Since the Moser spindle requires at least 4 colors and is a subgraph of the unit graph of the plane, clearly $\chi(\mathbb{R}^2) \geq 4$.

Moving to a preliminary upper bound, we get $\chi(\mathbb{R}^2) \leq 7$ via an unpublished coloring of the plane due to Isbell (according to Soifer [191, p. 24]), presented in Figure 4.9. We leave it to the reader as Exercise 4.4 to prove that this provides a 7-coloring of $\mathbb{R}^2$ with no two points at unit distance of the same color.

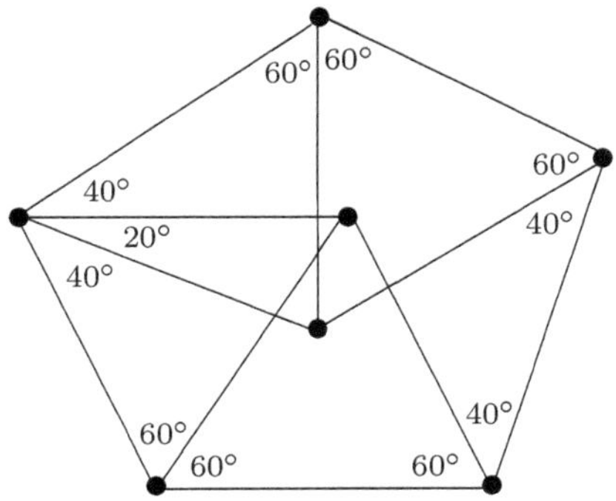

FIGURE 4.8
The Moser spindle proves $\chi(\mathbb{R}^2) \geq 4$; all edges have unit length

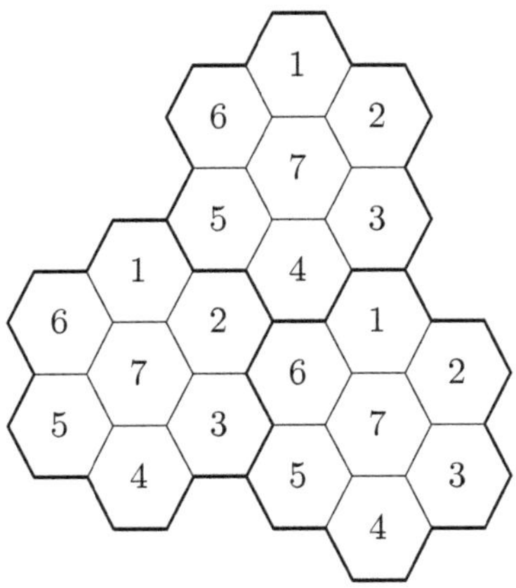

FIGURE 4.9
A 7-coloring used to cover $\mathbb{R}^2$ (the color pattern among the group of seven hexagons outlined in bold tiles the plane) with no two points at unit distance of the same color; the diameter of the regular hexagons is $\frac{15}{16}$

So far, we have

$$4 \leq \chi(\mathbb{R}^2) \leq 7.$$

Both bounds were discovered by 1950 and remained the best-known until 2018 when computer scientist and biologist de Grey [55] provided a unit graph on 1581 vertices that is not 4-colorable. Like many computational problems in graph-based Ramsey theory, much of the legwork is done before any computation occurs since brute-force algorithms do not have realistic stopping times. No progress on the upper bound has been achieved. Thus, we have the following theorem (see [73] and references therein for the higher dimension bounds).

Theorem 4.11. *The chromatic number of the plane is between 5 and 7, inclusive, i.e.,*

$$5 \leq \chi(\mathbb{R}^2) \leq 7.$$

Furthermore, $6 \leq \chi(\mathbb{R}^3) \leq 15$ *and* $9 \leq \chi(\mathbb{R}^4) \leq 54$.

Since de Grey's breakthrough announcement, the online math collective Polymath has worked to decrease the number of vertices needed to prove $\chi(\mathbb{R}^2) > 4$. As of this writing, the smallest such published graph has 553 vertices and is due to Heule [106] (this is the same Heule who helped prove the 2-regularity of Pythagorean triples, improved some bounds on classical van der Waerden numbers, and determined the Schur number $s(5) = 161$). According to Polymath members, the smallest number of vertices obtained as of this writing is 510, due to Jaan Parts.

4.3 Four Color Map Theorem

In the previous sections we focused on vertices of polygons. In this section our focus shifts slightly (but only to start with, as we will see). Consider the 6-sided die in Figure 3.5, reproduced below for ease of reading.

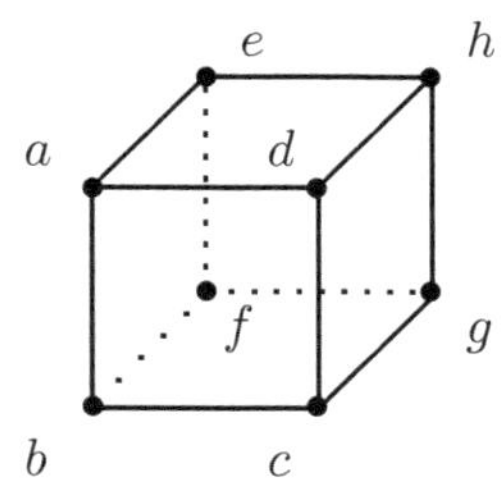

It is not difficult to determine that the vertices are 2-colorable (with color classes $\{a, c, f, h\}$ and $\{b, d, e, g\}$. Can we say anything about coloring the faces of the cube so that no two faces that share an edge have the same color? It is helpful to "flatten" out the cube to deal with a graph in $\mathbb{R}^2$. To do so, consider the sphere centered at the cube's center which contains the cube's vertices. Performing the stereographic projection onto a plane tangent to the sphere and parallel to the bottom face of the cube we obtain the graph in Figure 4.10.

The representation of a cube in Figure 4.10 is an example of a planar graph (defined next), and with planar graphs we have a method of viewing

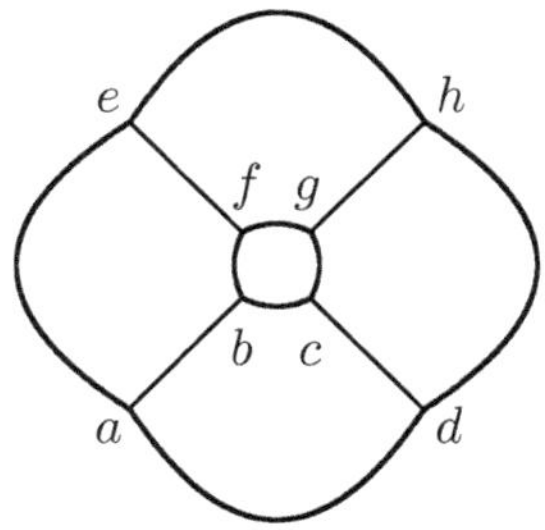

FIGURE 4.10
Stereographic projection of a cube

the coloring of faces as the coloring of vertices by looking at the dual graph. Note that the face $\{a, d, e, h\}$ on the cube is represented by the unbounded region "outside" of the figure in Figure 4.10.

Definition 4.12 (Planar graph, Dual of a planar graph)**.** A *planar graph* is a graph that can be drawn in $\mathbb{R}^2$ in such a way that no two edges intersect except, possibly, at a vertex. Given a planar graph, the *dual graph* is created by identifying each region (bounded and unbounded) with a vertex and connecting two vertices with an edge if and only if the two associated regions share an edge.

We present the dual of our cube in Figure 4.11.

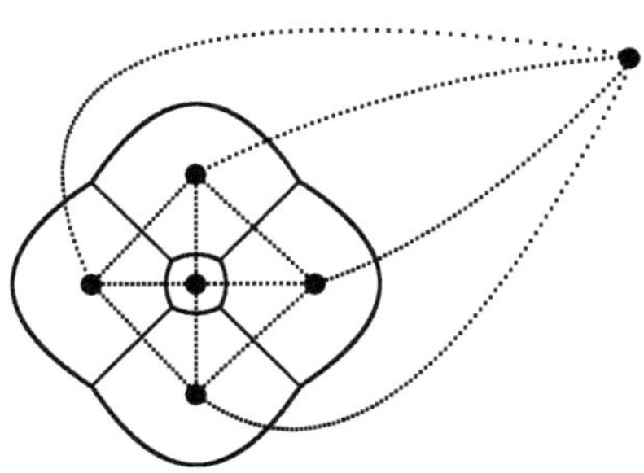

FIGURE 4.11
Dual graph of the stereographic projection of a cube

By construction, if the dual if r-colorable, then in the original planar graph we can color the regions with r colors and avoid two adjacent regions having the same color. In the case of the dual graph in Figure 4.11, the minimal number of colors needed for our dual graph to be r-colorable is easily determined to be $r = 3$. Hence, when using three colors we can color the faces of a cube and not have two adjacent faces of the same color (on the cube this is achieved by coloring opposite sides with the same color).

Notation. Let G be a planar graph. We denote by G^* the dual of G.

Recalling the definition of the chromatic number of a graph G (Definition 3.12), by considering G^* we see that the minimal number of colors needed to color the regions of G so that no two regions sharing a boundary have the same color is $\chi(G^*)$.

The question of interest in this section is to determine the maximum of $\chi(G^*)$ over all planar graphs G. This has been commonly referred to as determining the minimal number of colors needed to color the countries of a (geographical) map so that no two neighboring countries have the same color. In this phrasing there as some technicalities to address (e.g., countries need to be contiguous), so we will stick to the mathematical formulation. That is, letting $\mathcal{P}$ be the set of planar graphs, does

$$\max_{G\in\mathcal{P}} \chi(G^*)$$

exist, and if so what is its value?

We can easily determine that, if this maximum exists, it must be at least 4 by considering the planar graph in Figure 4.12 (a slight modification of the stereographic projection of the cube).

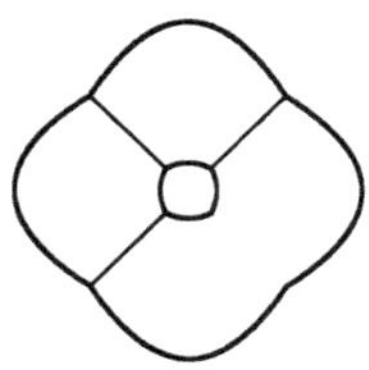

FIGURE 4.12
A planar graph requiring 4 colors

Based on an incomplete proof of Kempe [118], Heawood [103] provided the following upper bound.

Theorem 4.13 (Five Color Theorem). *Let $\mathcal{P}$ be the set of planar graphs. Then*

$$\max_{G\in\mathcal{P}} \chi(G^*) \leq 5.$$

Proof. We take as a fact not proven here that if G is a planar graph, then G^* is planar. We will also use Euler's Formula:

$$|V| - |E| + |F| = 2$$

for any planar graph $H = (V, E)$ with $|F|$ faces/regions.

We start by showing that H must have a vertex incident to at most 5 edges. Assume, for a contradiction, that this is false. Then each vertex has at least 6 edges incident to it. Summing over all vertices and noting that doing so double counts edges, we obtain $|E| \geq 3|V|$. Next, since we have assumed throughout this book that all graphs are simple, each face is bound by at least 3 edges. Summing over all faces and noting that this also double counts edges, we find that $2|E| \geq 3|F|$. Using these bounds in Euler's Formula we obtain

$$2 = |V| - |E| + |F| \leq \frac{|E|}{3} - |E| + \frac{2|E|}{3} = 0,$$

a contradiction.

Having this vertex degree bound, we prove the theorem by induction on the number of vertices in a dual of a planar graph, where the base case on 3 vertices is trivial. Let G^* be an arbitrary (dual of a) planar graph. Let v be a vertex of G^* with at most 5 edges incident to it. Delete v and the edges incident to it. The remaining graph G' is 5-colorable by the inductive assumption. If there are less than 5 colors used by vertices connected to v in G^*, then we can color v with any color not used by these vertices and inserting v and its edges back in, we remain 5-colorable. Hence, we are in the situation where v is connected to 5 vertices, say a, b, c, d, and e, using the colors $1, 2, 3, 4$, and 5, respectively. By deleting v and its incident edges, the remaining graph G' is 5-colorable. (Note that it may be that G' is not connected; however, this is considered below.)

Since planarity is a required condition, we must consider the orientation of the vertices we are considering. Hence, we may assume the orientation given in Figure 4.13.

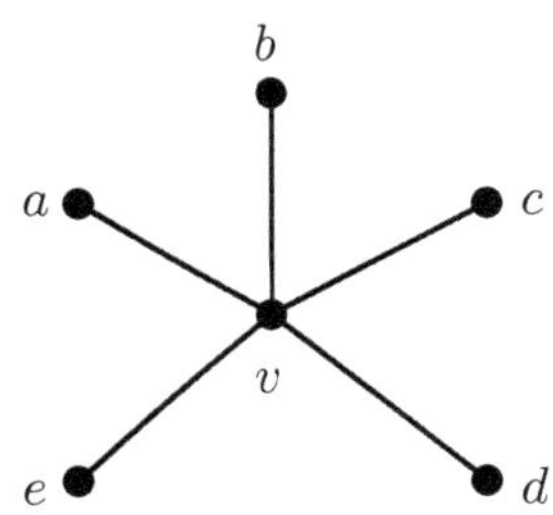

FIGURE 4.13
Orientation of vertices for the proof of the Five Color Theorem

Let $G'_{1,3}$ be the induced subgraph of G' consisting only of vertices colored 1 or 3 and the edges between them.

Consider a possible path in $G'_{1,3}$ from a to c traveling along vertices alternately colored 1 and 3 (such a path must alternate colors due to the colorability of G'). If no such path exists, then $G'_{1,3}$ must consist of at least two (separate) components. Interchange the colors in the component containing a. When G' is now considered with colors 1 and 3 switched, the coloring is still valid; however, both a and c have color 3 so that we may color v with color 1 to see that G^* is 5-colorable. Hence we may assume a path P_1 from a to c of alternating colors 1 and 3 exists.

Analogously considering $G'_{2,4}$ we may assume the existence of a path P_2 from b to d of alternating colors 2 and 4.

Now, by planarity, P_1 and P_2 cannot have an intersecting edge. This means that P_1 and P_2 must share a vertex. However, this vertex must be colored from both $\{1,3\}$ and $\{2,4\}$, which is not possible. Hence, both paths cannot exist. Since G^* was taken as an arbitrary (dual of a) planar graph, we conclude that $\max_{G\in\mathcal{P}} \chi(G^*) \leq 5$. □

Theorem 4.13 stood for 87 years until Appel and Haken (see [6] for the announcement; see [7, 8] for the proof) provided the first major theorem proved via computer. Because the proof was computer-assisted and the first significant such proof, many mathematicians found the argument unconvincing. However, over time it has come to be accepted and several other proofs and proof-checking techniques have confirmed the correctness of the result. As all known proofs are computer-assisted, we do not prove this theorem here.

Theorem 4.14 (Four Color Theorem). *Let $\mathcal{P}$ be the set of planar graphs. Then*

$$\max_{G\in\mathcal{P}} \chi(G^*) = 4.$$

4.4 Exercises

4.1 Consider a regular n-gon. You are allowed to insert k diagonals (i.e., an edge connecting any two non-adjacent vertices) but must not create a triangle consisting of three vertices from the n-gon. What is the maximum k can be as a function of n?

4.2 Let T be an equilateral triangle and let S be a unit square. Show that both $E(T, 2; 2)$ and $E(S, 2; 2)$ are false.

4.3 Show that the Moser spindle is not 3-colorable.

4.4 Prove that $\chi(\mathbb{R}^2) \leq 7$ by showing that the coloring given in Figure 4.9 does not admit two points at unit distance of the same color.

4.5 Let χ be a 2-coloring of $\mathbb{R}^2$ that contains a monochromatic equilateral triangle of side length $s \in \{3, 4, 5\}$. Prove that χ also contains a monochromatic right triangle (from first principles; i.e., not by appealing to a known result).

4.6 Prove that every 2-coloring of $\mathbb{Z}^2$ admits a monochromatic isosceles right triangle without appealing to any result in this chapter.

4.7 Consider the five Platonic solids: tetrahedron, cube, octahedron, dodecahedron, and icosahedron. By the Four Color Theorem, each requires at most 4 colors to color the faces so that no two faces sharing an edge have the same color. In fact, one of the five solids requires only 2 colors and only one of them requires 4 colors (the others require 3 colors). Classify the Platonic solids by the minimal number of required colors.

4.8 Consider the obvious generalization of planar maps to $\mathbb{R}^3$ (non-intersecting regions with positive volume). Prove that for any $r \in \mathbb{Z}^+$, there exists a 3-dimensional "map" that requires r colors if no two adjacent regions are allowed to be the same color. Deduce that there is no generalization of the Four Color Theorem to higher dimensions.

4.9 Let G be a planar graph. Prove that G is 2-colorable if and only if the degree of every vertex of G is even.

4.10 Let G^* be the dual graph of a planar graph with the property that every region is bound by at least 4 edges. Prove, without the Four Color Theorem, that G^* is 4-colorable. Strengthen this to show that if at most $\frac{2}{3}^{\text{rd}}$s of all regions are bound by only 3 edges then G^* is still 4-colorable.

5

Other Approaches to Ramsey Theory

It seems that the problem is very deep, 'cause every time I try to sleep I have nightmares.

–Gordon Gano

Many of the most recent results in Ramsey theory stem from the approaches presented in this chapter. We will briefly delve into these approaches by giving proofs of previously encountered results. Although we have attempted to produce content that is as self-contained as possible, some antecedent results in the areas of analysis, topology, dynamical systems, and abstract algebra are stated without proof and will be specialized to serve our purposes. We will, of course, point the reader to relevant references.

5.1 Topological Approaches

Topology can effectively address Ramsey theory questions due to its flexibility in defining spaces so that sets of colorings can be considered in a structured setting. Topology can also give us a way of defining "closeness" that can address how similarly colored two structures are. As we will see, compactness (see Definition 1.13) is a key component of the mix.

5.1.1 Proof of van der Waerden's Theorem

Consider the arithmetic progression $a, a+d, a+2d, \dots, a+(k-1)d$ as positions on the number line. To move from a to $a+d$ we would shift our position by d units. Of course this same shift moves us to the next term of the progression. This perspective sets us up to discuss dynamical systems.

Definition 5.1 (Topological dynamical system)**.** We call (X, T) a *topological* *dynamical system* if X is a compact topological metric space and $T : X \to X$ is a homeomorphism (i.e., a topological isomorphism; see Definition 1.19).

Recalling that a compact metric space is closed and bounded, it should make intuitive sense that if U is any open set in X, then by repeated application of T some point in U must eventually return to U. To see this, consider the pre-images of U, i.e., $T^{-1}(U), T^{-2}(U), \ldots$. Since T is a homeomorphism, these are open sets. They cover an open subset of X, which contains a closed non-empty subset (Let $a \in \bigcup_{i=1}^{\infty} T^{-i}(U)$ so that there exists $\epsilon > 0$ such that $\{x \in X : d(x,a) < \epsilon\} \subseteq A$. Then $\{x \in X : d(x,a) \leq \frac{\epsilon}{2}\} \subseteq A$ is closed.) Having an open cover of a closed (and hence, compact) subset, the cover must admit a finite subcover. This means that there must be intersections between them, i.e.,

$$T^{-n}(U) \cap T^{-m}(U) \neq \emptyset$$

for some $n > m$. Taking x in this intersection we see that both $T^n(x)$ and $T^m(x)$ are in U. Since $T^{n-m}(T^m(x)) = T^n(x) \in U$ we see that $T^m(x) \in U$ returns to U after $n - m$ repeated applications of T.

The following result extends this recurrence property and is crucial for our purpose. The theorem is due to Furstenberg and Weiss [78] but is referred to using Birkhoff's name since Birkhoff proved the "single" recurrence case [22].

Theorem 5.2 (Multiple Birkhoff Recurrence Theorem). *Let (X, T) be a topological dynamical system. Then for any $k \in \mathbb{Z}^+$, there exists $x \in X$ and a sequence $\{n_j\}_{j=1}^{\infty} \to \infty$ of positive integers such that*

$$\lim_{j \to \infty} T^{in_j}(x) = x$$

for all $i \in \{1, 2, \ldots, k\}$ (simultaneously).

At the beginning of this subsection we discussed shifts of integers. This will be the homeomorphism we use once we define the space X appropriately. We will then use Theorem 5.2 to deduce van der Waerden's Theorem.

Proof of van der Waerden's Theorem. We first define our topological space. Let X be the space of bi-directionally infinite sequences

$$\ldots x_{-2}, x_{-1}, x_0, x_1, x_2, \ldots$$

with $x_i \in \{0, 1, \ldots, r-1\}$ for all $i \in \mathbb{Z}$. For our transformation we take $T : X \to X$ to be the (left) shift operator, i.e., $T(\ldots x_{-2}, x_{-1}, x_0, x_1, x_2, \ldots)$ is defined as the sequence resulting from performing $x_i \to x_{i-1}$ for all $i \in \mathbb{Z}$.

To endow X with a metric, for $x = \{x_i\}_{i \in \mathbb{Z}} \neq y = \{y_i\}_{i \in \mathbb{Z}} \in X$, we first define

$$m(x, y) = \min(\{|i| : x_i \neq y_i\}).$$

Using this, we define the metric (the validity of this is left to the reader as Exercise 5.1):

$$d(x, y) = 2^{-m(x,y)} \quad \text{for } x \neq y, \quad \text{and } d(x, x) = 0. \tag{5.1}$$

In order to show that X is compact, since it is infinite dimensional we must show that X is complete and totally bounded (see Theorem 1.17). With this metric, X is clearly complete. To show that X is totally bounded we must show that for any $\epsilon > 0$, there exist open sets $U_1, U_2, \ldots, U_k$ of radius at most ϵ that cover X. To this end, given $\epsilon > 0$, determine $n \in \mathbb{Z}^+$ such that

$$\frac{1}{2^n} < \epsilon \leq \frac{1}{2^{n-1}}.$$

Consider all possible sequences of length n over $\{0, 1, \ldots, r-1\}$. For each of the r^n sequences $s_0^{(i)}, s_1^{(i)}, \ldots, s_{n-1}^{(i)}$, $1 \leq i \leq r^n$, let

$$s^{(i)} = s_{-(n-1)}^{(i)}, s_{-(n-2)}^{(i)} \ldots, s_{-1}^{(i)}, s_0^{(i)}, s_1^{(i)}, s_2^{(i)} \ldots, s_{n-1}^{(i)}$$

where $s_{-j}^{(i)} = s_j^{(i)}$.

Consider the sets

$$B_i = \{\ldots, x_{-(n+2)}, x_{-(n+1)}, x_{-n}, s^{(i)}, x_n, x_{n+1}, x_{n+2}, \ldots : 0 \leq x_j \leq r-1\}$$

for $1 \leq i \leq r^n$, and note that with $t^{(i)} = 0, 0, 0, \ldots, s^{(i)}, 0, 0, 0, \cdots \in X$, we can describe the B_i's by

$$B_i = \left\{ x \in X : d(x, t^{(i)}) < \epsilon \right\}.$$

To see that B_i is open, if $x \in B_i$ and $y \in X$ with $d(x, y) < \epsilon$ then we must have $d(y, t^{(i)}) < \epsilon$ so that $y \in B_i$. Since

$$\bigcup_{i=1}^{r^n} B_i = X$$

we see that X is totally bounded. This finishes the justification that X is compact.

We leave the proof that T is a homeomorphism to the reader as Exercise 5.2.

Now consider an arbitrary r-coloring $\chi : \mathbb{Z}^+ \to \{0, 1, \ldots, r-1\}$. We associate χ with the point

$$\ldots, 0, 0, 0, x_0, x_1, x_2, \ldots,$$

by defining $x_{i-1} = \chi(i)$ for all $i \in \mathbb{Z}^+$. Referring to this point by $x \in X$, consider

$$Y = \mathtt{cl}\left(\bigcup_{i=0}^{\infty} T^i(x) \right) \subseteq X$$

so that Y is a compact subspace with inherited metric d. Note further that $T : Y \to Y$ is still a homeomorphism.

By Theorem 5.2, there exists $y \in Y$ and $\{n_j\}_{j=1}^{\infty} \to \infty$ such that

$$T^{in_j}(y) \underset{j\to\infty}{\longrightarrow} y \quad \text{for all } 1 \le i \le k.$$

In particular, there exists $d \in \mathbb{Z}^+$ such that

$$d(T^{id}(y), y) < 1$$

for all $1 \le i \le k$. This metric assures us of agreement of colors of the "middle" terms of $T^{id}(y)$ and y. Hence, we can conclude that

$$y_0 = y_d = y_{2d} = \cdots = y_{kd}.$$

With k and d determined, find $a \in \mathbb{Z}^+$ such that

$$d(T^{a-1}(x), y) < 2^{-(kd+1)}$$

(existence is guaranteed by the definition of Y). This means that

$$x_{a-1} = y_0, x_{a-1+d} = y_d, \ldots, x_{a-1+kd} = y_{kd}$$

since we have color agreement of at least $kd+1$ terms between $T^{a-1}(x)$ and y. From this, we conclude that, under χ, the arithmetic progressions $a, a+d, a+2d, \ldots, a+kd$ is monochromatic, finishing this proof of van der Waerden's Theorem. □

5.1.2 Proof of the de Bruijn-Erdős Theorem

Recall Theorem 4.9, restated below for reference, which uses the chromatic number notation $\chi(G)$ (see Definition 3.23).

de Bruijn-Erdős Theorem. *Let G be a graph on infinitely many vertices. Let $\mathcal{H}$ be the family of finite subgraphs of G, i.e., of subgraphs with finite vertex sets. Assuming the Axiom of Choice, if $\chi(H) \le r$ for all $H \in \mathcal{H}$ with equality attained, then $\chi(G) = r$.*

In order to prove this in a topological setting, we need to define two topologies useful for graphs, along with a result equivalent to the Axiom of Choice.

Definition 5.3 (Discrete topology)**.** The *discrete topology* on X is the topology defined by taking every subset of X to be an open set.

Definition 5.4 (Product topology)**.** Let I be an index set and let $\{X_\alpha : \alpha \in I\}$ be a collection of sets. Let

$$X = \prod_{\alpha \in I} X_\alpha$$

be the Cartesian product (note that this notation extends the $A \times B$ notation used when $|I| < \infty$). The *product topology* on X is the topology obtained by defining the open sets of X to be

$$\left\{ \prod_{\alpha \in I} \mathcal{O}_\alpha : \mathcal{O}_\alpha \subseteq X_\alpha \text{ is open and } \mathcal{O}_\alpha = X_\alpha \text{ for all but finitely many } \alpha \right\}.$$

Without the restriction that $\mathcal{O}_\alpha = X_\alpha$ for all but finitely many α, the resulting topology is called the *box topology*. The box topology has many undesirable (for us) properties. In particular, even if all the X_α in $X = \prod_{\alpha \in I} X_\alpha$ are compact, X is not necessarily compact under the box topology. As we have seen in the last subsection, compactness is a crucial property for us. Fortunately, under the product topology, we have Tychonoff's Theorem (Theorem 5.5). Tychonoff's Theorem, which is equivalent to the Axiom of Choice, was proved in 1930 by Tychonoff [203] in the case of powers of $[0, 1]$ and in 1937 by Čech [38] in full generality (this is the same Čech whose name appears in the title of Section 5.3); see [216] for a modern treatment.

Theorem 5.5 (Tychonoff's Theorem). *Let I be an index set and let $\{X_\alpha : \alpha \in I\}$ be a collection of compact topological spaces. Then, under the product topology, the product $\prod_{\alpha \in I} X_\alpha$ is compact.*

How do we use these concepts to frame r-colorings of the vertices of an infinite graph G? First, let the vertices of the graph be indexed by I. For every $\alpha \in I$, let Y_α be the discrete topology defined on $\{0, 1, \ldots, r-1\}$. Clearly each Y_α is compact (there are only finitely many open sets for a given number of colors r). Defining

$$X = \prod_{\alpha \in I} Y_\alpha$$

equipped with the product topology, by Tychonoff's Theorem we see that X is compact and any coloring of the vertices of G can be viewed as a point in X. In this light, we now prove the de Bruijn-Erdős Theorem.

Proof of the de Bruijn-Erdős Theorem. Given an infinite graph $G = (V, E)$, define

$$X = \prod_{\alpha \in V} Y_\alpha,$$

where $Y = Y_\alpha$ is the space defined by the discrete topology on $\{0, 1, \ldots, r-1\}$ for each $\alpha \in V$, and equip X with the product topology. We view the points in X as r-colorings of the vertices of G.

For every $\{a, b\} \in E$, let

$$C_{\{a,b\}} = \{\gamma \in X : \gamma(a) \neq \gamma(b)\}.$$

We will show that $C_{\{a,b\}}$ is closed by showing that its complement

$$X \setminus C_{\{a,b\}} = \bigcup_{i=0}^{r-1} \{\chi \in X : \chi(a) = \chi(b) = i\}$$

is open. Since this is a union, it suffices to show that

$$\{\gamma \in X : \gamma(a) = \gamma(b) = c\}$$

is open for any $c \in \{0, 1, \ldots, r-1\}$. This set is (essentially)

$$\{c\} \times \{c\} \times \prod_{\alpha \in V \setminus \{a,b\}} Y. \tag{5.2}$$

By definition of the discrete topology, $\{c\}$ is an open subset of Y; hence, by the definition of product topology, the set in Expression (5.2) is open. This proves that each $C_{\{a,b\}}$ is a closed subset of X.

Consider

$$S = \bigcap_{(a,b) \in E} C_{\{a,b\}}.$$

Notice that any coloring in S has the property that any two vertices of G sharing an edge have different colors (i.e., such a coloring shows G is r-colorable). The proof will be complete upon showing that $S \neq \emptyset$.

Since S is the intersection of closed sets, we see that S is closed. We also know, via Tychonoff's Theorem, that X is compact.

By hypothesis, every finite subgraph of G is r-colorable so that any intersection of finitely many $C_{\{a,b\}}$'s is non-empty. However, we want to show that the infinite intersection describing S is non-empty.

Assume, for a contradiction, that $S = \emptyset$. Then $X \setminus S = X$ so that

$$X = \bigcup_{(a,b) \in E} X \setminus C_{\{a,b\}}.$$

Hence, we have a cover of X by open sets. Since X is compact, this admits a finite subcover, say

$$X = \bigcup_{i=1}^{k} X \setminus C_{\{a_i,b_i\}}.$$

Taking the complement of both sides we find that

$$\emptyset = \bigcap_{i=1}^{k} C_{\{a_i,b_i\}},$$

contradicting the hypothesis that such finite intersections are non-empty. We conclude that $S \neq \emptyset$ so that G is r-colorable, i.e., $\chi(G) \leq r$. Since we must use r colors for at least one finite graph, we have $\chi(G) \geq r$, and conclude that $\chi(G) = r$. □

5.1.3 Proof of Hindman's Theorem

Consider the approach to the proof of van der Waerden's Theorem given in Section 5.1.1. The structure of an arithmetic progression correlates perfectly with the shift operator. In turns out that this basic approach can be adapted (albeit with quite a bit more preliminary work) to tackle Hindman's Theorem, which we restate here for reference.

Hindman's Theorem. *Let $r \in \mathbb{Z}^+$. Every r-coloring of $\mathbb{Z}^+$ admits an infinite set $S = \{s_i\}_{i=1}^{\infty} \subseteq \mathbb{Z}^+$ such that*

$$FS(S) = \left\{ \sum_{i \in I} s_i : \emptyset \neq I \subseteq \mathbb{Z}^+, \, |I| < \infty \right\}$$

is monochromatic.

As we saw in Section 2.4.2, the combinatorial proof is quite involved. The topological one we present here is also quite involved. The topological proof is originally due to Furstenberg and Weiss [78]. The proof presented here follows the presentation of Setyawan [187].

As we will have need to appeal to Tychonoff's Theorem (Theorem 5.5), we assume the Axiom of Choice for this proof of Hindman's Theorem.

We will need more preliminaries in the realm of topological dynamical systems. We start with a couple of definitions.

Definition 5.6 (Proximal)**.** Let (X, T) be a topological dynamical system with metric d. We say that $x, y \in X$ are *proximal* if there exists a sequence of positive integers $\{n_j\}_{j=1}^{\infty} \to \infty$ such that

$$\lim_{j \to \infty} d(T^{n_j}(x), T^{n_j}(y)) = 0.$$

Within the framework used for van der Waerden's Theorem's proof, the points in X were viewed as colorings. From this viewpoint, two colorings are proximal (under the metric given in Equation (5.1)) if there are arbitrarily long intervals of color agreement.

Definition 5.7 (Function space)**.** Let X be a compact space and let $F := \{f : X \to X\}$ be the set of functions on X. This set can be described by points in

$$X^X = \prod_{\alpha \in X} X_\alpha$$

where $X_\alpha = X$ for all $\alpha \in X$ since $g \in X^X$ has its βth coordinate equal to $g(\beta)$.

We will be using the function space to investigate the behavior of repeated applications of the shift transform T. By applying Tychonoff's Theorem, we immediately see that X^X is a compact space under the product topology.

We now present a lemma, credited to Ellis [62] and Numakura [153], that is not only crucial for this section, but also for Section 5.3. It proves the existence of idempotents in certain compact spaces. (We take the concept of right-continuous to mean that applying the binary operation from the right is continuous.)

Lemma 5.8 (Ellis-Numakura Lemma). *Let $(S, *)$ be a non-empty compact Hausdorff semigroup with $*$ being right-continuous (or left-continuous). Assuming the Axiom of Choice, S contains an idempotent, i.e., there exists $p \in S$ such that $p * p = p$.*

Proof. Let

$$\mathcal{F} = \{A \subseteq S : A * A \subseteq A,\ A \text{ is compact},\ A \neq \emptyset\}.$$

Since $S \in \mathcal{F}$ we see that $\mathcal{F} \neq \emptyset$. Consider the poset of $\mathcal{F}$ under set inclusion. Let $\{A_t : t \in T\}$ be a chain, with $A_i \supseteq A_j$ for $i < j$, in this poset and define

$$B = \bigcap_{t \in T} A_t.$$

We will show that $B \neq \emptyset$ (this is true by Cantor's Intersection Theorem, but we will prove it here).

Assume, for a contradiction, that $B = \emptyset$. Let

$$U_t = S \setminus A_t$$

for all $t \in T$ and note that U_t is open for all $t \in T$. Furthermore, we see that

$$\bigcup_{t \in T} U_t = S \setminus B = S,$$

so that we have an open cover of S. Since S is compact, this cover admits a finite subcover $U_{t_1} \subseteq U_{t_2} \subseteq \cdots \subseteq U_{t_k}$ of nested sets, since $A_{t_1} \supseteq A_{t_2} \supseteq \cdots \supseteq A_{t_k}$. This means that $U_{t_k} = S$, which implies that $A_{t_k} = \emptyset$, contradicting the definition of $\mathcal{F}$.

Since B is compact, non-empty, and $B * B \subseteq B$ by construction, we see that $B \in \mathcal{F}$. Hence, every chain $\{A_t : t \in T\}$ of $\mathcal{F}$ has a lower bound (namely, B) in $\mathcal{F}$. Applying Zorn's Lemma (Lemma 4.10), we find that $\mathcal{F}$ has a minimal element, say M. Let $p \in M$ and consider $M * p$. Since $M \in \mathcal{F}$ we have $M * p \subseteq M$.

We are given that operating on the right by p is continuous. Since continuous functions map compact sets to compact sets and we know that M is compact, we see that $M * p$ is compact. Hence, by the minimality of M we can conclude that $M * p = M$.

We can now conclude that there exists $h \in M$ such that $h * p = p$. Let

$$H = \{m \in M : m * p = p\}.$$

Since $h \in H$, we see that $H \neq \emptyset$. Now, since S is a semigroup, if $h_1, h_2 \in H$, we have

$$(h_1 * h_2) * p = h_1 * (h_2 * p) = h_1 * p = p$$

so that $h_1 * h_2 \in H$. Thus, $H * H \subseteq H$. In order to conclude that $H \in \mathcal{F}$, we must show that H is compact.

Clearly $H \subseteq M * p$. Since closed subsets of compact sets are compact, and $M * p$ is compact, it suffices to show that H is closed.

Let P be the pre-image of p under operation on the right by p. Then $H = M \cap P$. We know that M is compact, and hence closed. Since our map is right-continuous and $\{p\}$ is a closed set (by the Hausdorff assumption), we see that P is closed. As H is the intersection of two closed sets, H is closed.

We can now conclude that $H \in \mathcal{F}$. Since $H \subseteq M$, by the minimality of M we have $H = M$. Thus, $p \in H$, so we have $p * p = p$. □

Remark. All topological spaces we consider are Hausdorff.

Before getting to the proof of Hindman's Theorem, we still need a few more definitions and preliminary results.

Definition 5.9 (Minimal)**.** We say that a topological dynamical system (X, T) is *minimal* if $T(Y) \subseteq Y$, with $\emptyset \neq Y \subseteq X$ and Y compact, occurs only for $Y = X$.

Definition 5.10 (Uniformly recurrent)**.** Let (X, T) be a topological dynamical system. We say that $x \in X$ is *uniformly recurrent* if for all $\epsilon > 0$, there exists $N \in \mathbb{Z}^+$ such that for any $a \in \mathbb{Z}^+ \cup \{0\}$, there exists $n \in [a+1, a+N]$ such that $d(T^n(x), x) < \epsilon$.

The preliminary results we will rely on are next.

Theorem 5.11. *Let (X, T) be a topological dynamical system. Then there exists (a compact set) $Y \subseteq X$ such that $(Y, T|_Y)$ is minimal.*

Theorem 5.12. *If (X, T) is a minimal topological dynamical system, then each $x \in X$ is uniformly recurrent.*

We will not prove Theorems 5.11 and 5.12, but will provide a proof of the next theorem as it is the crux of the proof of Hindman's Theorem.

Theorem 5.13. *Let (X, T) be a topological dynamical system. For any $x \in X$, there exists a uniformly recurrent point $y \in X$ such that x and y are proximal.*

Proof. Take $x \in X$ and let

$$X_x = \mathtt{cl}\left(\{T^n(x) : n \in \mathbb{Z}^+ \cup \{0\}\}\right).$$

Then X_x is a closed subset of a compact space and, hence, is compact. View

(X_x, T) as a topological dynamical system (which is why we note that X_x is compact). By Theorem 5.11, there exists

$$Y \subseteq X_x$$

such that $(Y, T|_Y)$ is minimal.

Define E to be the closure

$$E = \texttt{cl}\left(\{T^n : n \in \mathbb{Z}\}\right)$$

in the function space $X_x^{X_x}$ (under the product topology) and define

$$F = \{f \in E : f(x) \in Y\}.$$

`Claim.` F is a non-empty, compact semigroup that is right-continuous under composition.

We first show that $F \neq \emptyset$. Let $w \in Y$. Then $w \in X_x$ so that for all $j \in \mathbb{Z}^+$ there exists $n_j \in \mathbb{Z}^+$ such that $d(w, T^{n_j}(x)) < 2^{-j}$. Since $\{T^{n_j}(x)\}_{j \in \mathbb{Z}^+}$ is an infinite sequence in a compact metric space, let $g(x)$ be a subsequential limit and note that $g \in E$. Then

$$d(g(x), w) \leq d(w, T^{n_{j_\ell}}(x)) + d(T^{n_{j_\ell}}(x), g(x)) < 2^{-j_\ell} + 2^{-j_\ell}$$

for infinitely many $j_\ell \in \mathbb{Z}^+$. Hence, $d(g(x), w) = 0$ so that $g(x) = w \in Y$, giving us $g \in F$.

To see that F is compact, it suffices to show that F is closed as F is a subset of a compact space. Let $\{g_n\}_{n \in \mathbb{Z}^+} \subseteq F$ with $\lim_{n \to \infty} g_n = g$. Hence, $g_n(x) \in Y$ for all $n \in \mathbb{Z}^+$. We know that Y is closed so that $\lim_{n \to \infty} g_n(x) = h(x) \in Y$ for some function h. Under the product topology, which is also known as the topology of pointwise convergence, we have $h(x) = g(x)$ (note that this does not mean $h = g$). Hence, $g \in F$ so that F is closed, and hence, compact.

Next we will show that F is a semigroup under composition of functions (associativity is standard). Let $f, g \in F$. We must show that $f \circ g \in F$. Since $f \in E$, there exists a sequence $\{n_j\}_{j \in \mathbb{Z}^+}$ such that $d(T^{n_j}(g(x)), f(g(x)) < 2^{-j}$ for all $j \in \mathbb{Z}^+$. Since Y is a compact metric space, there exists a subsequence of $\{n_j\}_{j \in \mathbb{Z}^+}$ such that $\lim_{\ell \to \infty} T^{n_{j_\ell}}(g(x)) = v$ for some $v \in Y$. Hence, we can conclude that $d(T^{n_{j_\ell}}(g(x)), v) < 2^{-j_\ell}$ for infinitely many $j_\ell \in \mathbb{Z}^+$. Thus,

$$d(f(g(x)), v) \leq d(f(g(x)), T^{n_{j_\ell}}(g(x))) + d(T^{n_{j_\ell}}(g(x)), v) < 2^{-j_\ell} + 2^{-j_\ell}$$

for infinitely many $j_\ell \in \mathbb{Z}^+$. Hence, $f(g(x)) = v \in Y$ so that $f \circ g \in F$.

Lastly, we show that F is right-continuous under composition. Let $\{f_i\}_{i \in \mathbb{Z}^+} \subseteq F$ converge to $f \in F$ so that for any $t \in Y$ we have that $\{f_i(t)\}_{i \in \mathbb{Z}^+}$ converges to $f(t)$. Consider $g \in F$ so that $g(x) \in Y$. Letting $t = g(x)$ we see that $\{f_i(g(x))\}_{i \in \mathbb{Z}^+}$ converges to $f(g(x))$, giving us right-continuity.

This concludes the claim's justification. $\diamond$

We now apply Lemma 5.8 to find the existence of an idempotent in F (F

is Hausdorff by noting that it inherits the metric from Y: we have $d(f,g) = d(f(x), g(x))$ since both $f(x)$ and $g(x)$ are in Y by definition of F). Let p be this idempotent. With $x \in X$ still fixed, let $y = p(x) \in Y$. By Theorem 5.12, y is uniformly recurrent. We proceed by showing that x and y are proximal.

We have $p \in E$. In order to finish the proof, we go back to what the open sets in the product topology are. Loosely speaking, they are finite products of open sets of X_x. Since $p \in E$, $x \in X_x$, and $y = p(x) \in Y \subseteq X_x$ we see that for any $j \in \mathbb{Z}^+$ there exists $n_j \in \mathbb{Z}^+$ such that

$$d(T^{n_j}(x), p(x)) = d(T^{n_j}(x), y) < 2^{-j}$$

and

$$d(T^{n_j}(p(x)), (p \circ p)(x)) = d(T^{n_j}(y), p(x)) = d(T^{n_j}(y), y) < 2^{-j}$$

both hold. From this, we see that

$$d(T^{n_j}(x), T^{n_j}(y)) \leq d(T^{n_j}(x), y) + d(T^{n_j}(y), y) \leq 2^{-j} + 2^{-j} = 2^{-j+1}$$

for all $j \in \mathbb{Z}^+$, and we conclude that x and y are proximal. □

With these preliminaries under our belt, we are now in a position to prove Hindman's Theorem.

Proof of Hindman's Theorem. Let χ be an r-coloring of $\mathbb{Z}^+$. Associate this with the point $\ldots, \chi(3), \chi(2), \chi(1), \chi(2), \chi(3), \ldots \in X$, where X is the space of bi-directionally infinite sequences $\ldots x_{-2}, x_{-1}, x_0, x_1, x_2, \ldots$ with $x_i \in \{0, 1, \ldots, r-1\}$ for all $i \in \mathbb{Z}$. Let T be the shift operator and let d be the metric given in Equation (5.1). With this metric, note that T is continuous. Letting

$$X_\chi = \mathtt{cl}\left(\{T^n(\chi) : n \in \mathbb{Z}^+ \cup \{0\}\}\right)$$

we see that (X_χ, T) is a topological dynamical system.

By Theorem 5.13 there exists a uniformly recurrent point γ with

$$\gamma \in X_\chi \text{ such that } \chi \text{ and } \gamma \text{ are proximal.}$$

Our goal is to find an infinite set $\{m_i\}_{i \in \mathbb{Z}^+}$ such that

$$d(T^s(\chi), \gamma) < 1$$

for any finite sum

$$s \in FS(\{m_i\}_{i \in \mathbb{Z}^+})$$

so that under our metric, $\chi(s)$ has the same color as γ_0 (the "middle" color of γ) for any such s, meaning that

$$\{\chi(s) : s \in FS(\{m_i\}_{i \in \mathbb{Z}^+})\}$$

contains only the color γ_0.

We summarize the preliminary results from this subsection below, as they pertain to our situation. Through iterating a series of implication that these results give us, we will be able to determine the existence of an appropriate sumset.

(UR) `Uniform Recurrence.` For any $\epsilon > 0$, there exists $N \in \mathbb{Z}^+$ such that for any $a \in \mathbb{Z}^+$ there exists $k_1 \in [1, N]$ such that

$$d(T^{a+k_1}(\gamma), \gamma) < \frac{\epsilon}{2}.$$

(UC) `Uniform Continuity.` Given $\epsilon > 0$ and $N \in \mathbb{Z}^+$, for each $i \in [1, N]$ there exists $\delta_i > 0$ such that for any $x, y \in X$ if $d(x, y) < \delta_i$ then

$$d(T^i(x), T^i(y)) < \frac{\epsilon}{2}.$$

(Note that T^i is continuous on a compact space so that it is uniformly continuous.)

(P) `Proximal.` Given $\delta_i > 0$ for $1 \leq i \leq N$, there exists $k_2 \in \mathbb{Z}^+$ such that

$$d(T^{k_2}(\chi), T^{k_2}(\gamma)) < \min_{i \in [1,N]} \delta_i.$$

Here is how we use these results. Let $\epsilon > 0$ be given. Apply (UR) to determine N. Use ϵ and this N in (UC) to get the δ_i's. Use the δ_i's in (P) to get k_2. From the definition of proximal, we may take $k_2 > N$ so that $k_2 > k_1$ in the following step. With ϵ and N still the same, use $a = k_2$ in (UR) to get k_1. Again with the same ϵ and N, apply (UC) with

$$x = T^{k_2}(\chi), \quad y = T^{k_2}(\gamma), \quad \text{and} \quad i = k_1.$$

Using the triangle inequality now gives us $d(T^{k_1+k_2}(\chi), \gamma) < \epsilon$ and we already have $d(T^{k_1+k_2}(\gamma), \gamma) < \frac{\epsilon}{2} < \epsilon$. Hence, we can conclude that for any $\epsilon > 0$, there exist (infinitely many) $m \in \mathbb{Z}^+$ such that

$$d(T^m(\chi), \gamma) < \epsilon \quad \text{and} \quad d(T^m(\gamma), \gamma) < \epsilon.$$

Now, given $\epsilon_1 > 0$, let $m_1 = k_1 + k_2$ be as determined above with $\epsilon = \frac{\epsilon_1}{4}$. Choose $N > m_1$ and take ϵ_2 sufficiently small to apply (UC) with $\delta_{m_1} = \epsilon_2 < \frac{\epsilon_1}{2}$ and $\epsilon = \frac{\epsilon_1}{2}$ so that

$$d(T^{m_1}(\tau), T^{m_1}(\gamma)) < \frac{\epsilon_1}{4}$$

for any coloring τ with $d(\tau, \gamma) < \epsilon_2$. Then

$$d(T^{m_1}(\tau), \gamma) \leq d(T^{m_1}(\tau), T^{m_1}(\gamma)) + d(T^{m_1}(\gamma), \gamma) \leq \frac{\epsilon_1}{4} + \frac{\epsilon_1}{4} = \frac{\epsilon_1}{2}. \quad (5.3)$$

Now choose $m_2 > m_1$ so that

$$d(T^{m_2}(\chi), \gamma) < \epsilon_2 \quad \text{and} \quad d(T^{m_2}(\gamma), \gamma) < \epsilon_2.$$

Using $\tau = T^{m_2}(\chi)$ in Inequality (5.3), we obtain

$$d(T^{m_1+m_2}(\chi), \gamma) < \frac{\epsilon_1}{2}$$

so that

$$\begin{aligned} d(T^{m_1+m_2}(\gamma), \gamma) &< d(T^{m_1+m_2}(\gamma), T^{m_1+m_2}(\chi)) + d(T^{m_1+m_2}(\chi), \gamma) \\ &< \frac{\epsilon_1}{2} + \frac{\epsilon_1}{2} = \epsilon_1. \end{aligned}$$

Hence, we have shown that if $d(T^{m_1}(\chi), \gamma)$ and $d(T^{m_1}(\gamma), \gamma)$ are both less than ϵ then there exists $m_2 > m_1$ such that $d(T^{m_2}(\chi), \gamma) < \frac{\epsilon}{4}$, and $d(T^{m_1+m_2}(\chi), \gamma)$ and $d(T^{m_1+m_2}(\gamma), \gamma)$ are both less than ϵ. (Taking $\epsilon = 1$ is all that is needed to provide perhaps the most difficult proof of Schur's Theorem available.)

We now just repeat this argument. Formally, we continue inductively and let $M = \{m_i\}_{i=1}^{k}$ be such that for all $s \in FS(M)$ we have $d(T^s(\chi), \gamma) < \epsilon$ and $d(T^s(\gamma), \gamma) < \epsilon$. Let $s \in FS(M)$ be arbitrary and apply the above argument to find $m_{k+1} > m_k$ with $d(T^{m_{k+1}}(\chi), \gamma) < \frac{\epsilon}{4}$, $d(T^{s+m_{k+1}}(\chi), \gamma) < \epsilon$, and $d(T^{s+m_{k+1}}(\gamma), \gamma) < \epsilon$, thereby completing the induction.

Hence, by taking $\epsilon = 1$, we have shown the existence of $m_1, m_2, \ldots$ such that for any $s \in FS(\{m_i\}_{i=1}^{\infty})$ we have $d(T^s(\chi), \gamma) < 1$ as needed to prove Hindman's Theorem. □

5.2 Ergodic Theory

In last section's proof of Hindman's Theorem, a fair bit of analysis entered the picture. Even further in this direction, we have measure-theoretic dynamical systems. The Krylov-Bogoliubov Theorem [127], stated below, shows the connection between the topological dynamical systems and the measure-theoretic framework we will investigate.

Theorem 5.14 (Krylov-Bogoliubov Theorem). *Let X be a compact, topological metric space with continuous map $T : X \to X$. Let $\mathcal{B}$ be the Borel σ-algebra generated by the family of open subsets of X. Then there exists a probability measure $\mu : \mathcal{B} \to [0,1]$ such that $\mu(A) = \mu(T^{-1}(A))$ for any $A \in \mathcal{B}$.*

In essence, this theorem gives us a way of treating the compact metric spaces we use in the topology setting as finite spaces with a well-behaved measure.

Notation. We use $(X, \mathcal{B}, \mu)$ to represent a measure space where $\mathcal{B}$ is the family of measurable sets of X and μ is the measure.

Definition 5.15 (Measure-preserving map)**.** A map $T : (X, \mathcal{B}_1, \mu) \to (Y, \mathcal{B}_2, \nu)$ is *measure-preserving* if for any $A \in \mathcal{B}_2$ we have $\mu(T^{-1}(A)) = \nu(A)$.

Instead of topological dynamical systems, we will be interested in measure-preserving dynamical systems. We specialize our definition to probability spaces as all results in this section concern probability spaces.

Definition 5.16 (Measure-preserving dynamical system)**.** We call $(X, \mathcal{B}, T, \mu)$ with $\mu(X) = 1$ a *measure-preserving dynamical system* if $T : (X, \mathcal{B}, \mu) \to (X, \mathcal{B}, \mu)$ is measure-preserving.

For our purpose, we will still be considering the space X of bi-directionally infinite sequences with terms from $\{0, 1, \ldots, r-1\}$, along with the shift operator. For $B \subseteq X$, we let B_j be the jth component and define

$$b_j = \frac{|B_j|}{r}$$

(i.e., the uniform probability measure on the finite set $\{0, 1, \ldots, r-1\}$). Let

$$\mu(B) = \prod_{j \in \mathbb{Z}} b_j.$$

Under the shift operator T, we see that

$$T(B_i) = B_{i-1}$$

so the system is clearly measure-preserving.

Quite a bit of machinery from ergodic theory is needed to present Furstenberg's proof of Szemerédi's Theorem (Theorem 2.70) in a self-contained manner. We refer the reader to the very lucid write-up by Wang [213] for such details. Our goal here is to present some of the main results and ideas needed for the proof, attempting to show similarities with the ideas presented in the last section.

The first question you may have at this point is: What is ergodic theory? In short, it is the study of ergodic systems:

Definition 5.17 (Ergodic system)**.** We say that a measure-preserving dynamical system $(X, \mathcal{B}, T, \mu)$ is *ergodic* if for any $A \in \mathcal{B}$ satisfying $T^{-1}(A) = A$ we have either $\mu(A) = 1$ or $\mu(A) = 0$.

Compare this definition with that of a minimal topological dynamical system (Definition 5.9). Noting that we are dealing with measure-preserving systems, an ergodic system is perfectly analogous to a minimal system, up to measure 0.

We can also give an intuitive notion of what an ergodic system is by considering a bucket of white paint and a cup of blue tint. We take as our measure the volume, so that, letting W be the white paint and B be the blue tint, we have $\mu(W \sqcup B) = 1$ and both $\mu(W)$ and $\mu(B)$ are non-zero. We let our transformation T be some manner of stirring the paint and tint together. As a painter, our goal is to have the white paint and blue tint mixed to a point that the color is uniform throughout the bucket. If our stirring T is chosen in such a way that some nontrivial region in the bucket is not touched, then we will not reach our goal. On the other extreme, we don't need T to achieve uniform mixture by itself, but we do hope that repeated stirrings, i.e., application of T^j, will get us there.

With μ as our volume measure, uniform concentration means that for some $j \in \mathbb{Z}^+$,

$$\frac{\mu(A \cap T^j(B))}{\mu(A)} = \mu(B),$$

is satisfied for any region A in the bucket, since $\mu(B)$ is the fraction of blue tint in the entire bucket (our total volume is 1).

Since T is measure-preserving (we don't lose any volume by stirring) we apply T^{-j} (yes, we can't physically un-stir to reverse the process) to get

$$\mu(T^{-j}(A \cap T^j(B))) = \mu(T^{-j}(A) \cap B).$$

So, our goal is to have

$$\mu(T^{-j}(A) \cap B) = \mu(A)\mu(B) \tag{5.4}$$

for any region A as this would give us a uniform concentration throughout the bucket.

Since we are dealing with analysis, we know that showing Equation (5.4) holds for some fixed j is too much to ask for. We can, however, ask that

$$\lim_{j \to \infty} \mu(T^{-j}(A) \cap B) = \mu(A)\mu(B).$$

When this occurs, we say that our system is *strongly mixing*.

A weaker concept, which would still allow for a fairly uniform looking paint mixture, is allowing Equation (5.4) to occur on average:

$$\lim_{n \to \infty} \frac{1}{n} \sum_{j=0}^{n-1} \left| \mu(T^{-j}(A) \cap B) - \mu(A)\mu(B) \right| = 0.$$

If this occurs, we say that our system is *weakly mixing*. (For the interested reader, Chacon [39] has provided a measure-preserving map that is weakly, but not strongly, mixing.)

We now come to the even weaker concept of *ergodic*. We only ask now that after some finite number of applications of our stirring transformation, regions interact, i.e., there exists $m \in \mathbb{Z}^+$ such that

$$\mu(T^{-m}(A) \cap B) > 0.$$

This means that our stirring moves some blue tint arbitrarily close to any point infinitely many times. Physically, our bucket has many different shades of blue, but no region is pure white and no region is pure blue. In terms of what this is compared to weakly mixing, it can be shown that ergodic systems satisfy the following:

$$\lim_{n\to\infty} \frac{\sum_{j=0}^{n-1} \mu(T^{-j}(A) \cap B)}{n} = \mu(A)\mu(B).$$

There are several stronger concepts of mixing, but the study of mixing in dynamical systems is referred to as ergodic theory since ergodicity is the weakest concept needed to prove many classical results in the area. Hopefully the above paint-mixing example gives a bit more information on the types of systems being considered.

5.2.1 Furstenberg's Proof of Szemerédi's Theorem

The goal of this section is to give the reader an idea of the method of proof used by Furstenberg [76] to prove Szemerédi's Theorem, which we state below for reference.

Szemerédi's Theorem. *For any $k \in \mathbb{Z}^+$ and any subset $A \subseteq \mathbb{Z}^+$, if $\limsup_{n\to\infty} \frac{|A\cap[1,n]|}{n} > 0$, then A contains a k-term arithmetic progression.*

Remark. The upper density requirement given in our statement of Szemerédi's Theorem can be replaced by the broader conditions that $A \subseteq \mathbb{Z}$ and

$$\limsup_{m\to\infty} \frac{|A \cap [-m, m]|}{2m+1} > 0.$$

This density is called the *upper Banach density.*

The concepts involved with Furstenberg's proof are weakly mixing and compactness and are used to extend Poincaré's Recurrence Theorem, which was, according to Oxtoby [154], proved (mostly) by Poincaré (see [159]), and given a modern treatment by Carathéodory [36].

Theorem 5.18 (Poincaré Recurrence Theorem). *Let $(X, \mathcal{B}, T, \mu)$ be a measure-preserving dynamical system. Let $A \in \mathcal{B}$. Then for almost every $x \in A$ (i.e., up to a set of measure zero) we have*

$$|\{n : T^n(x) \in A\}| = \infty;$$

in other words, x returns to A infinitely often.

Comparing this recurrence theorem with the recurrence theorems in Section 5.1, you may suspect that Furstenberg's extension has to do with making the recurrence more uniform, or regular. Indeed, this is where we are headed, but first we prove Poincaré's Recurrence Theorem.

Proof of Theorem 5.18. Our goal is to prove the existence of $B \subseteq A$ with $\mu(A \setminus B) = 0$ so that for all $b \in B$ we have $T^n(b) \in A$ for infinitely many n. To this end, naïvely define

$$B = \{b \in A : T^n(b) \in A \text{ for infinitely many } n \in \mathbb{Z}^+\}.$$

For $k \in \mathbb{Z}^+ \cup \{0\}$, define

$$A_k = \{a \in A : T^k(a) \in A \text{ but } T^n(a) \notin A \text{ for all } n > k\}.$$

Note that $A_k = T^{-k}(A_0)$.

We have

$$B = A \setminus \bigcup_{k=0}^{\infty} A_k,$$

so by showing that

$$\mu\left(\bigcup_{k=0}^{\infty} A_k\right) = 0$$

we will be done. This reduces to showing that $\mu(A_k) = 0$ for all $k \in \mathbb{Z}^+ \cup \{0\}$ since the countable union of measure-zero sets has measure zero.

For $0 \leq i < j$, we have

$$A_i \cap A_j = T^{-i}(A_0) \cap T^{-j}(A_0) = T^{-i}(A_0 \cap T^{-j+i}(A_0)).$$

We will show that $A_i \cap A_j = \emptyset$ (even though this should be clear by definition). To this end, assume, for a contradiction, that there exists

$$x \in A_0 \cap T^{-j+i}(A_0).$$

Since $x \in A_0$, we have $T^n(x) \notin A$ for all $n \in \mathbb{Z}^+$, but $x \in T^{-j+i}(A_0)$ means that $T^{j-i}(x) \in A_0 \subseteq A$, a contradiction.

We now see that for $i \neq j$, the sets $T^{-i}(A_0)$ and $T^{-j}(A_0)$ are disjoint. Since our system is measure-preserving, we know that

$$\mu(T^{-k}(A_0)) = \mu(A_0)$$

for all $k \in \mathbb{Z}^+ \cup \{0\}$.

We can now conclude that

$$\mu\left(\bigsqcup_{k=0}^{\infty} T^{-k}(A_0)\right) = \sum_{k=0}^{\infty} \mu(T^{-k}(A_0)) = \sum_{i=0}^{\infty} \mu(A_0).$$

Since our entire space has total measure 1, we must have $\mu(A_0) = 0$. We conclude that
$$\mu(A_k) = \mu(T^{-k}(A_0)) = \mu(A_0) = 0$$
for any $k \in \mathbb{Z}^+ \cup \{0\}$.

As noted before, we have $A_{k_1} \cap A_{k_2} = \emptyset$ for $k_1 \neq k_2$. Hence,
$$\mu\left(\bigsqcup_{k=0}^{\infty} A_k\right) = \sum_{k=0}^{\infty} \mu(A_k) = \sum_{k=0}^{\infty} 0 = 0,$$
and we conclude that since
$$A = B \sqcup \left(\bigsqcup_{k=0}^{\infty} A_k\right)$$
we have the existence of B with $\mu(A \setminus B) = 0$ having the property that for all $b \in B$ we have $T^n(b) \in A$ for infinitely many n. □

As we did in the last section, we will consider the space X of bi-directionally infinite sequences along with the shift operator. However, we are not coloring integers this time, but, given a set $S \subseteq \mathbb{Z}^+$ we can view this as a 2-coloring of $\mathbb{Z}^+$ by letting the terms in the infinite sequence be defined by $x_{-i} = x_i = 1$ if $i \in S$ and 0 if $i \notin S$. Hence, we can use the framework provided in the proof of Hindman's Theorem in Section 5.1.3 (with $r = 2$). We will use the notation $X = \{0,1\}^{\mathbb{Z}}$.

Instead of appealing to Tychonoff's Theorem (Theorem 5.5) as we did in Section 5.1.3, we will use the following easy lemma, the proof of which is left to the reader as Exercise 5.6.

Lemma 5.19. *The set of probability measures in $\{0,1\}^{\mathbb{Z}}$ is closed.*

The careful reader may notice that none of the results in this subsection have any type of mixing hypothesis. This is because the heavy lifting for the proof of Szemerédi's Theorem is done with Furstenberg's Multiple Recurrence Theorem where quite a bit of ergodic machinery is used to prove this result. The reader is referred to Furstenberg's original article [76] as well as [213] for details as we will not offer a proof.

Theorem 5.20 (Furstenberg's Multiple Recurrence Theorem)**.** *Let $(X, \mathcal{B}, T, \mu)$ be a measure-preserving dynamical system. For any $A \in \mathcal{B}$ with $\mu(A) > 0$ and any $k \in \mathbb{Z}^+$, there exists $d \in \mathbb{Z}^+$ such that*
$$\mu(A \cap T^{-d}(A) \cap T^{-2d}(A) \cap \cdots \cap T^{-(k-1)d}(A)) > 0.$$

We will deduce Szemerédi's Theorem assuming Theorem 5.20.

Proof of Szemerédi's Theorem. Let $X = \{0,1\}^{\mathbb{Z}}$ and let $A \subseteq \mathbb{Z}$ be represented by $x = \dots x_{-1}x_0x_1 \dots \in X$ with $x_i = 1$ if $i \in A$ and $x_i = 0$ if $i \notin A$. Consider the set $B = \texttt{cl}(\{T^n(x) : n \in \mathbb{Z}\})$ and let $C \subseteq B$ be those non-accumulation points whose coordinate in position 0 (the "middle" position) has value 1.

We will prove the theorem as described in the remark following the theorem's statement as given at the beginning of this subsection.

Consider the probability measures for $m \in \mathbb{Z}^+$ given by

$$\mu_m(S) = \frac{1}{2m+1} \sum_{i=-m}^{m} \delta_{T^i(x)}(S),$$

where $\delta_t(S) = 1$ if $t \in S$ and 0 otherwise. Notice that

$$\mu_m(C) = \frac{|A \cap [-m, m]|}{2m+1}.$$

Let

$$\mu = \limsup_{m\to\infty} \mu_m.$$

By Lemma 5.19, μ is a probability measure on X.

Let T be the shift operator. We next show that T is measure-preserving on B under μ. Let $D \subseteq B$. Then

$$\begin{aligned}
|\mu(T^{-1}(D)) - \mu(D)| &= \limsup_{m\to\infty} \frac{1}{2m+1} \left| \sum_{i=-m}^{m} \delta_{T^i(x)}(T^{-1}(D)) - \sum_{i=-m}^{m} \delta_{T^i(x)}(D) \right| \\
&= \limsup_{m\to\infty} \frac{1}{2m+1} \left| \sum_{i=-m-1}^{m-1} \delta_{T^i(x)}(D) - \sum_{i=-m}^{m} \delta_{T^i(x)}(D) \right| \\
&= \limsup_{m\to\infty} \frac{1}{2m+1} \left| \delta_{T^{-m-1}(x)}(D) - \delta_{T^m(x)}(D) \right| \\
&\leq \limsup_{m\to\infty} \frac{1}{2m+1} = 0,
\end{aligned}$$

so that $\mu(T^{-1}(D)) = \mu(D)$.

We now have that $(B, \mathcal{B}_B, T, \mu)$ (using the induced sigma-algebra for B) is a measure-preserving dynamical system in addition to having a set $C \subseteq B$ with $\mu(C) > 0$. Applying Theorem 5.20, we have $d \in \mathbb{Z}^+$ such that

$$\mu(C \cap T^{-d}(C) \cap T^{-2d}(C) \cap \cdots \cap T^{-(k-1)d}(C)) > 0.$$

Since μ is a supremum limit, there exists $M \in \mathbb{Z}^+$ such that

$$\mu_M(C \cap T^{-d}(C) \cap T^{-2d}(C) \cap \cdots \cap T^{-(k-1)d}(C)) > 0.$$

Thus, there exists $c \in [-M, M]$ such that

$$c \in C \cap T^{-d}(C) \cap T^{-2d}(C) \cap \cdots \cap T^{-(k-1)d}(C).$$

Now, by definition of C, we have $c = T^a(x)$ for some $a \in \mathbb{Z}$. Since the coordinate in position 0 of c has value 1, we have $a \in A$. Similarly, $c \in T^{-jd}(C)$ means that $a + jd \in A$. Thus, we see that $a, a+d, a+2d, \ldots, a+(k-1)d$ are all members of A, thereby finishing the proof. □

Remark. If we are restricted to A as a subset of $\mathbb{Z}^+$, note that if the upper density of A is at least ϵ then the upper Banach density of A is at least $\frac{\epsilon}{2}$ so that we can conclude that Szemerédi's Theorem as stated at the beginning of this subsection holds (note that C in the above proof contains only $T^a(x)$ with $a \in \mathbb{Z}^+$ if $A \subseteq \mathbb{Z}^+$).

5.3 Stone-Čech Compactification

We now change gears and move away from dynamical systems to an idea borne from topology. As we have seen previously, compactness plays a crucial role in proving Ramsey-type theorems via topological dynamical systems. If our natural space X (e.g., $X = \mathbb{Z}^+$) is not compact, we would like to turn it into a compact space Y such that X is a dense subset of Y. As a trivial example, consider the intervals $X = (0,1)$ and $Y = [0,1]$.

There are different methods to construct the compactification of a general topological space X. We will be concerned only with $X = \mathbb{Z}^+$ and will use a set-based approach.

Before getting to this, a simple example is in order. Consider $\mathbb{R}$ with its usual topology. This is not compact, but consider $Y = \mathbb{R} \cup \{\infty\}$, i.e, add the "point at infinity." The topology of Y is the set of open sets of $\mathbb{R}$ and all sets of the form $U \cup \{\infty\}$, where U is an open subset of $\mathbb{R}$ with the restriction that $\mathbb{R} \setminus U$ is compact. For example, one such U is $(-\infty, -1) \cup (1, \infty)$.

Now consider the embedding of $\mathbb{R}$ into the unit circle $S^1 \subseteq \mathbb{C}$ by $f(x) = e^{2i\arctan(x)}$. Then f is a bijection onto $S^1 \setminus \{-1\}$. Letting $f(\infty) = -1$ extends to a bijection between $\mathbb{R} \cup \{\infty\}$ and S^1. This is the one-point compactification of $\mathbb{R}$.

Our job in this section is more difficult as we want to compactify $\mathbb{Z}^+$. We know that $\mathbb{Z}^+$ is not compact in the (usual) discrete topology since $\bigcup_{i=1}^{\infty} \{i\}$ is an open cover of $\mathbb{Z}^+$ without any finite subcover. As a first step, let's find the one-point compactification of $\mathbb{Z}^+$. Consider the obvious mapping of $\mathbb{Z}^+ \cup \{\infty\}$ to

$$Y = \{0\} \cup \bigcup_{n \in \mathbb{Z}^+} \left\{ \frac{1}{n} \right\}.$$

The basic open sets of Y are $\{\frac{1}{n}\}$ along with sets of the form

$$\{0\} \cup \bigcup_{n \in \mathbb{Z}^+ \setminus F} \left\{ \frac{1}{n} \right\}, \tag{5.5}$$

where F is a finite subset of $\mathbb{Z}^+$. Any open cover of Y must include one of the sets in (5.5) in order to cover 0, and hence, must admit a finite subcover, showing that Y is compact.

An issue with this one-point compactification of $\mathbb{Z}^+$ is in the name: we've only added one more point. How much more information can come of this? Furthermore, if we are hoping to gain information from the space containing all limit points, we see that every increasing sequence has the same limit point, namely 0 (in the compactification). Hence, we lose a lot of information under this compactification. We want a finer compactification, where compactification C_1 is finer than compactification C_2 (both of the same space X) if there exists a surjective continuous map $f : C_1 \to C_2$ where $f|_X$ is the identity.

The one-point compactification is the coarsest compactification. The finest compactification should hopefully provide us with more useful information.

Definition 5.21 (Stone-Čech compactification)**.** The *Stone-Čech compactification of* $\mathbb{Z}^+$ is the finest compactification of $\mathbb{Z}^+$ and is denoted $\beta\mathbb{Z}^+$.

Since we are interested in arithmetic properties of $\mathbb{Z}^+$, our goal should be a framework with some algebraic properties. There are several descriptions of the Stone-Čech compactification: the most algebraic one uses ultrafilters.

The precise history of ultrafilters appears a bit nebulous. The names Reisz [167], Ulam [204], Tarski [198], and Cartan [37] are cited in various places. Their works are mixtures of analysis and topology, so ultrafilters seem to be a good fit as both analysis and topology have proven effective thus far.

In this section we will be concerned with the set $S = \mathbb{Z}^+$ in the following definition.

Definition 5.22 (Filter, Ultrafilter on S)**.** Let p be a family of subsets of a set S. If p satisfies all of:

(i) $\emptyset \notin p$;

(ii) if $A \in p$ and $A \subseteq B$, then $B \in p$;

(iii) if $A, B \in p$, then $A \cap B \in p$,

then we say that p is a *filter on* S. If, in addition, p satisfies

(iv) for any $C \subseteq S$ either $C \in p$ or $S \setminus C \in p$,

then we say that p is an *ultrafilter on* S.

Remark. Item (iv) means that p is not properly contained in any other filter.

The family $\{A \subseteq \mathbb{Z}^+ : |\mathbb{Z}^+ \setminus A| < \infty\}$ is a filter, known as the Fréchet filter, but not an ultrafilter. For any fixed $n \in \mathbb{Z}^+$, the family $\{A \subseteq \mathbb{Z}^+ : n \in A\}$ is an ultrafilter and these will serve our purpose.

Definition 5.23 (Principle ultrafilter)**.** Let $n \in \mathbb{Z}^+$. We call (and denote by)

$$p_n = \{A \subseteq \mathbb{Z}^+ : n \in A\}$$

the *principle ultrafilter corresponding to* n.

Taking $f : \mathbb{Z}^+ \to \beta\mathbb{Z}^+$ as $f(n) = p_n$ gives an embedding of $\mathbb{Z}^+$ into $\beta\mathbb{Z}^+$. This mapping serves as the means to compactify $\mathbb{Z}^+$. The interested reader is pointed to the thorough book by Hindman and Strauss [111] for details. We will assume that $\beta\mathbb{Z}^+$ is a compactification of $\mathbb{Z}^+$, that

$$\beta\mathbb{Z}^+ = \{p : p \text{ is an ultrafilter of } \mathbb{Z}^+\},$$

and that

$$\{S_\beta : S \subseteq \mathbb{Z}^+\}, \quad \text{where} \quad S_\beta = \{p \in \beta\mathbb{Z}^+ : S \in p\}$$

is a basis for the topology of $\beta\mathbb{Z}^+$ consisting of open (and closed) sets S_β and defines a compact Hausdorff space.

Now that we have a compact space, what do we do if we have abandoned dynamical systems? To see how ultrafilters can be useful for Ramsey-type problems, consider the following lemma.

Lemma 5.24. *Let $r \in \mathbb{Z}^+$ and let p be an ultrafilter on set S. For any r-coloring of S, one of the color classes is in p.*

Proof. Let

$$S = \bigsqcup_{i=1}^{r} C_i,$$

where C_i is the ith color class. Assume, for a contradiction, that none of the color classes is in p. By part (iv) of Definition 5.22, we have $S \setminus C_i \in p$ for $1 \le i \le r$. Repeatedly applying part (iii) of Definition 5.22 we obtain

$$\bigcap_{i=1}^{r-1} (S \setminus C_i) \in p.$$

But

$$\bigcap_{i=1}^{r-1} (S \setminus C_i) = S \setminus \bigcup_{i=1}^{r-1} C_i = C_r$$

so $C_r \in p$, a contradiction. □

5.3.1 Proof of Schur's and Hindman's Theorems

This section is based on Glazer's proof of Hindman's Theorem, which was given (with permission) by Comfort in [48]. Along the way we will prove Schur's Theorem as a precursor to Hindman's Theorem as both results concern monochromatic sums of integers. In Section 5.1.3, the dynamical system we used to prove Hindman's Theorem relied on a shift map. We no longer have a shift map: as a substitute, we have the following arithmetic properties of $\beta\mathbb{Z}^+$ that relate to standard arithmetic in $\mathbb{Z}^+$.

Definition 5.25 (Addition in $\boldsymbol{\beta\mathbb{Z}^+}$)**.** Let $A \subseteq \mathbb{Z}^+$ and let

$$A - n = \{m \in \mathbb{Z}^+ : m + n \in A\}.$$

We define the *addition* $\oplus$ *of two ultrafilters* p and q in $\beta\mathbb{Z}^+$ by

$$A \in p \oplus q \quad \text{if and only if} \quad \{n \in \mathbb{Z}^+ : A - n \in p\} \in q.$$

Notice that the definition of addition is based on shifts of sets. This is the mechanism which takes the place of the shift map.

We leave the proof of the following lemma to the reader as Exercise 5.8.

Lemma 5.26. *With $\oplus$ as in Definition 5.25, $(\beta\mathbb{Z}^+, \oplus)$ is a semigroup.*

We will also need to appeal to the following lemma. For the proof, recall that $S_\beta = \{p \in \beta\mathbb{Z}^+ : S \in p\}$ is a basis for the topology of $\beta\mathbb{Z}^+$.

Lemma 5.27. *Let $q \in \beta\mathbb{Z}^+$ be fixed and define $f_q : \beta\mathbb{Z}^+ \to \beta\mathbb{Z}^+$ by $f_q(x) = q \oplus x$. Then f_q is continuous.*

Proof. This is essentially an unravelling of definitions. Taking a neighborhood S_β of $q \oplus x$ we want to show that there exists a neighborhood T_β of x such that whenever x' is in T_β we have that $q \oplus x'$ is in S_β. Letting S and T be the subsets of $\mathbb{Z}^+$ corresponding to S_β and T_β, respectively, we take $T = \{n : S - n \in q\}$ from the definition of $\oplus$. Since $S \in q \oplus x$ we have that $T \in x$. We want to show that if $T \in x'$ then $S \in q \oplus x'$. This is true by the definition of $\oplus$. □

Using Lemma 5.27 and appealing to Lemma 5.8, we obtain the result which allows us to prove the theorems in this section, namely, the existence of an idempotent (which was also used in the topological dynamical system proof of Hindman's Theorem; see Section 5.1.3).

Corollary 5.28. *There exists $p \in \beta\mathbb{Z}^+$ such that $p \oplus p = p$.*

We are now ready to delve into the proofs. For reference, here are the statements of Schur's and Hindman's theorems.

Schur's Theorem. *Let $r \in \mathbb{Z}^+$. Every r-coloring of $\mathbb{Z}^+$ admits a monochromatic solution to $x + y = z$.*

Hindman's Theorem. *Let* $r \in \mathbb{Z}^+$. *Every* r*-coloring of* $\mathbb{Z}^+$ *admits a set* $S = \{s_i\}_{i=1}^{\infty} \subseteq \mathbb{Z}^+$ *such that*

$$FS(S) = \left\{ \sum_{i \in I} s_i : \emptyset \neq I \subseteq \mathbb{Z}^+, \ |I| < \infty \right\}$$

is monochromatic.

Proof of Schur's Theorem. Let $p \in \beta\mathbb{Z}^+$ be an idempotent. For any r-coloring of $\mathbb{Z}^+$, by Lemma 5.24 one of the color classes, call it A, is in p. Hence, $A \in p = p \oplus p$. This means that

$$B = \{n \in \mathbb{Z}^+ : A - n \in p\} \in p.$$

By part (iii) of Definition 5.22, we have $A \cap B \in p$. Take

$$d \in A \cap B.$$

Then $d \in B$ gives $A - d \in p$. Take

$$a \in A \cap (A - d) \in p$$

(since $\emptyset \notin p$ by part (i) of Definition 5.22, we know $A \cap (A - d) \neq \emptyset$). Hence, we have $a, d, a + d \in A$ and we are done. □

Essentially, the proof of Schur's Theorem given above is the first step toward proving Hindman's Theorem via ultrafilters. However, we must find an infinite set instead of a set with three terms. Looking at the proof of Schur's Theorem above, all we needed was for $A \cap (A - d) \neq \emptyset$. We need much more now; we need this to be infinite. Toward this end, we will rely on the following lemma.

Lemma 5.29. *Let* $p \in \beta\mathbb{Z}^+$ *be an idempotent. Then* p *does not contain a finite set.*

Proof. We will first show that if $q \in \beta\mathbb{Z}^+$ contains a finite set, then q is a principle ultrafilter. Let $A \in q$ be a finite set and let $a \in A$. By part (iv) of Definition 5.22 either $\{a\} \in q$ or $\mathbb{Z}^+ \setminus \{a\} \in q$. If $\{a\} \in q$, then the principle ultrafilter corresponding to a is in q. If $\mathbb{Z}^+ \setminus \{a\} \in q$ then by intersecting this with A we can conclude that $A \setminus \{a\} \in q$. In this situation, we repeat the process with $A \setminus \{a\}$ in place of A. Since A if finite, this process will eventually end so that we can conclude for some $a' \in A$ the principle ultrafilter corresponding to a' is in q.

To show that q is itself a principle ultrafilter we assume, for a contradiction, that it is not (but that it contains one). Let $B \in q$ such that $B \notin p_a$ (the principle ultrafilter corresponding to a). Then $\mathbb{Z}^+ \setminus B \in p_a \subseteq q$. Hence, we have both B and $\mathbb{Z}^+ \setminus B$ in q. Since $B \cap (\mathbb{Z}^+ \setminus B) = \emptyset$, we have $\emptyset \in q$, contradicting the definition of a filter.

To finish the proof, we will show that if $p \in \beta\mathbb{Z}^+$ is an idempotent, then p cannot be a principle ultrafilter. For a contradiction, assume otherwise; that is, assume that $p = p_t$ for some $t \in \mathbb{Z}^+$. Let $A \in p \oplus p$ so that

$$\{n \in \mathbb{Z}^+ : A - n \in p\} \in p.$$

Since $p = p_t$ we have that $t \in A - t$ so that $2t \in A$. As this holds for every set A we see that

$$p_{2t} \supseteq p_t \oplus p_t.$$

We will now show that we have $p_{2t} = p_t \oplus p_t$. Assume, for a contradiction, that there exists $B \in p_{2t}$ such that $B \notin p_t \oplus p_t$. Then $\mathbb{Z}^+ \setminus B \in p_t \oplus p_t$. By definition of $\oplus$, we see that

$$C = \{n \in \mathbb{Z}^+ : (\mathbb{Z}^+ \setminus B) - n \in p_t\} \in p_t.$$

Since $C \in p_t$ we have $t \in C$. Thus

$$(\mathbb{Z}^+ \setminus B) - t \in p_t$$

so that $t \in (\mathbb{Z}^+ \setminus B) - t$. We conclude that $2t \in \mathbb{Z}^+ \setminus B$. This is a contradiction, since we have $B \in p_{2t}$ meaning $2t \in B$.

Since $p_t \oplus p_t = p_{2t} \neq p_t$, we have shown that if p is an idempotent, then it cannot be a principle ultrafilter. Since every ultrafilter that contains a finite set must be a principle ultrafilter, we see that p cannot contain a finite set. □

We now have the last necessary tool we need to prove Hindman's Theorem via ultrafilters.

Proof of Hindman's Theorem. Let $p \in \beta\mathbb{Z}^+$ be an idempotent. For any r-coloring of $\mathbb{Z}^+$, by Lemma 5.24 one of the color classes, call it A, is in p. We will show that A contains an infinite set $S \subseteq \mathbb{Z}^+$ such that $FS(S) \subseteq A$, which proves the theorem.

For any $T \subseteq \mathbb{Z}^+$, define

$$B_T = \{n \in \mathbb{Z}^+ : T - n \in p\}.$$

Since $A \in p \oplus p$ we have $B_A \in p$. By part (iii) of Definition 5.22, we have $A \cap B_A \in p$. From Lemma 5.29, we know that $A \cap B_A$ is infinite.

We construct S inductively. Let $A_1 = A$ and let $s_1 \in A_1 \cap B_{A_1}$ so that $A_1 - s_1 \in p$. Define

$$A_2 = A_1 \cap (A_1 - s_1) \in p$$

and note that by Lemma 5.29 we have that A_2 is also infinite. Then $A_2 \cap B_{A_2} \in p$ so there exists $s_2 > s_1$ such that $s_2 \in A_2 \cap B_{A_2}$. Hence, $A_2 - s_2 \in p$. Continue in this fashion to obtain

$$s_i \in A_i \cap B_{A_i} \text{ where } A_i = A_{i-1} \cap B_{A_{i-1}}.$$

Let $S = \{s_i\}_{i \in \mathbb{Z}^+}$ and note that $A = A_1 \supseteq A_2 \supseteq A_3 \supseteq \cdots$. We will show that $FS(S) \subseteq A$.

Let $\{j_1, j_2, \ldots, j_k\} \subseteq \mathbb{Z}^+$ be arbitrary with $j_1 < j_2 < \cdots < j_k$. We will show that

$$\sum_{i=1}^{k} s_{j_i} \in A_{j_1}$$

by induction on k. For $k = 1$ the result is clear since $s_i \in A_i$ for all $i \in \mathbb{Z}^+$.

Hence, we assume

$$s = \sum_{i=2}^{k} s_{j_i} \in A_{j_2}$$

by the inductive assumption. Since $A_{j_1} \supseteq A_{j_1+1} \supseteq A_{j_2}$ we have $s \in A_{j_1+1}$. By construction,

$$A_{j_1+1} = A_{j_1} \cap (A_{j_1} - s_{j_1}).$$

Hence, $s \in A_{j_1} - s_{j_1}$ so that

$$s_{j_1} + s = \sum_{i=1}^{k} s_{j_i} \in A_{j_1},$$

completing the induction.

Since $A_{j_1} \subseteq A$ for any $j_1 \in \mathbb{Z}^+$, we see that $FS(S) \subseteq A$. □

5.3.2 Proof of the de Bruijn-Erdős Theorem

Recall the de Bruijn-Erdős Theorem from Section 4.2:

de Bruijn-Erdős Theorem. *Let G be a graph on infinitely many vertices. Let $\mathcal{H}$ be the family of finite subgraphs of G, i.e., of subgraphs with finite vertex sets. Assuming the Axiom of Choice, if $\chi(H) \leq r$ for all $H \in \mathcal{H}$ with equality attained, then $\chi(G) = r$.*

In 1962, Luxemburg [139] provided an ultrafilter-based proof of this result. We follow this approach as conveyed by Lambie-Hanson [128].

We will use, without proof, the fact that every filter is contained in an ultrafilter (this follows from Zorn's Lemma and the remark after Definition 5.22).

Proof of de Bruijn-Erdős Theorem. By removing the empty graph, we take $\mathcal{H}$ to be the set of all non-empty, finite subgraphs of G, and we may assume that $H \in \mathcal{H}$ means that H has a finite number of vertices but inherits *all* edges from G. For W a finite set of vertices, let $\mathcal{H}_W$ be those graphs in $\mathcal{H}$ that contain all vertices in W. If $W = \{v\}$ we write $\mathcal{H}_v$.

Define the family

$$\mathcal{F} = \{\mathcal{S} \subseteq \mathcal{H} : \mathcal{H}_W \subseteq \mathcal{S} \text{ for some finite set of vertices } W\}.$$

We will now show that $\mathcal{F}$ is a filter on $\mathcal{H}$. Referring to parts (i)-(iii) in Definition 5.22, we have $\emptyset \notin \mathcal{F}$ by definition of $\mathcal{H}$ so that (i) is satisfied. Next, let $\mathcal{S}_1, \mathcal{S}_2 \in \mathcal{F}$ with $\mathcal{S}_1 \subseteq \mathcal{S}_2$. Since there exists a finite set of vertices W such that $\mathcal{H}_W \subseteq \mathcal{S}_1$ we see that $\mathcal{H}_W \subseteq \mathcal{S}_2$ so that $\mathcal{S}_2 \in \mathcal{F}$. Hence, $\mathcal{F}$ satisfies part (ii). For part (iii), let $\mathcal{S}_1, \mathcal{S}_2 \in \mathcal{F}$. Then there exist finite vertex sets W_1 and W_2 such that $\mathcal{H}_{W_1} \subseteq \mathcal{S}_1$ and $\mathcal{H}_{W_2} \subseteq \mathcal{S}_2$. We have

$$\mathcal{H}_{W_1 \cup W_2} = \mathcal{H}_{W_1} \cap \mathcal{H}_{W_2} \subseteq \mathcal{S}_1 \cap \mathcal{S}_2$$

so that $\mathcal{S}_1 \cap \mathcal{S}_2 \in \mathcal{F}$, since $W_1 \cup W_2$ is finite. Hence, $\mathcal{F}$ satisfies part (iii) and we can conclude that $\mathcal{F}$ is a filter.

Having shown that $\mathcal{F}$ is a filter, there exists an ultrafilter $\mathcal{U}$ with $\mathcal{F} \subseteq \mathcal{U}$.

Fix a vertex v of G. We know that for any $H \in \mathcal{H}_v$ there exists a vertex-valid r-coloring of H. Call this vertex-valid coloring $\chi_H : H \to \{0, 1, \ldots, r-1\}$ and define

$$A_v(i) = \{H \in \mathcal{H}_v : \chi_H(v) = i\},$$

i.e., $A_v(i)$ is the set of finite graphs that include vertex v with a vertex-valid coloring assigning color i to v. Clearly

$$\bigsqcup_{i=0}^{r-1} A_v(i) = \mathcal{H}_v;$$

that is, we have an r-coloring of the elements of $\mathcal{H}_v \in \mathcal{F} \subseteq \mathcal{U}$.

We now claim that for some $c \in \{0, 1, \ldots, r-1\}$ we have $A_v(c) \in \mathcal{U}$. Assuming otherwise, by the definition of an ultrafilter, we have $\mathcal{H} \setminus A_v(i) \in \mathcal{U}$ for each color i. Hence,

$$\bigcap_{i=0}^{r-1} (\mathcal{H} \setminus A_v(i)) = \mathcal{H} \setminus \bigcup_{i=0}^{r-1} A_v(i) = \mathcal{H} \setminus \mathcal{H}_v \in \mathcal{U}.$$

But having $\mathcal{H}_v \in \mathcal{U}$ and $\mathcal{H} \setminus \mathcal{H}_v \in \mathcal{U}$ means that $\mathcal{H}_v \cap (\mathcal{H} \setminus \mathcal{H}_v) = \emptyset \in \mathcal{U}$, contradicting the definition of a filter. Hence, we can conclude that for every vertex v, there exists a color $c_v \in \{0, 1, \ldots, r-1\}$ such that $A_v(c_v) \in \mathcal{U}$.

We use these $A_v(c_v) \in \mathcal{U}$ to define a coloring of the vertices of G: let $\chi(v) = c_v$. The proof will be completed by showing that χ is a vertex-valid r-coloring of G.

Assume, for a contradiction, that under this coloring there exist vertices u and w connected by an edge in G with $\chi(u) = \chi(w) = j$ for some color j. We have $A_u(j), A_w(j) \in \mathcal{U}$ so that $A_u(j) \cap A_w(j) \in \mathcal{U}$. Let

$$B \in A_u(j) \cap A_w(j).$$

This means there is a vertex-valid r-coloring γ of a finite subgraph $B \in \mathcal{H}$ that contains both u and w and assigns color j to both u and w. As B inherits all edges from G, we have that u and w are connected by an edge in B. This means that γ is not a vertex-valid r-coloring of B, a contradiction.

Hence, we have shown the existence of a vertex-valid r-coloring of the vertices of G, as required. □

5.4 Additive Combinatorics Methods

As we saw in Section 2.5, Fourier analysis – and analysis, in general – can prove beneficial for solving some Ramsey-type problems. The general area in which analytic methods are used to attack combinatorial problems is called additive combinatorics. The general aim is a deep understanding of the simple operations of addition and multiplication.

As the theory is developed, it is a deeply analytic approach to combinatorics involving harmonic and Fourier analysis, probability, analytic number theory, and, of course, combinatorics. For an in-depth understanding of this approach, the reader is referred to the highly informative book by Tao and Vu [197]. Shorter introductions are given by Granville [86] and Green [87].

Many of the recent groundbreaking results in Ramsey Theory (e.g., Gowers' proof [82] of a new upper bound on van der Waerden numbers and Green and Tao's proof [90] that the primes contain arbitrary long arithmetic progressions), are realized via additive combinatorics.

You may wonder if we are just extending the approach to Roth's Theorem (Section 2.5.1). In some sense, the answer is yes; however, there is an obstacle to overcome. The issue is with the "small" Fourier coefficients. In the proof of Roth's Theorem (Theorem 2.68), if all of the Fourier coefficients (except $r = 0$) are small, we can conclude that the combined contribution from the small coefficients is not enough to cancel out the main term.

However, when dealing with longer arithmetic progressions, as noted by Soundararajan [192], the small Fourier coefficients may not lead to a "small" total contribution: For any $\delta > 0$, let $A \subseteq \mathbb{Z}_n$ where $a \in A$ if and only if there exists b with $|b| < \frac{1}{2}\delta n$ such that $a^2 \equiv b \pmod{n}$. Clearly $|A| \approx \delta n$. It turns out that the Fourier coefficients for the indicator function of A are all small (except for $\widehat{A}(0)$) but that A has significantly more 4-term arithmetic progressions than a random set. This poses a significant problem when trying to use Roth's proof with longer arithmetic progressions. Hence, advanced techniques are needed.

5.4.1 The Circle Method: Infinitely Many 3-term Arithmetic Progressions Among the Primes

The circle method can be considered the entry price to additive combinatorics. It was introduced by Hardy and Littlewood [100] with modifications by Vinogradov [210]. Conceptual work for the circle method was laid by Hardy and Ramanujan in [101].

The basic approach is to develop a generating function for counting the objects of interest and using integration on the complex unit circle. The circle is split into "major" and "minor" arcs, depending on their contributions to the complete integral. Since exact integration is not possible, good estimates are needed to complete the circle method.

In this section we attempt to explain the approach and tools used to show that the primes contain infinitely many 3-term arithmetic progressions. We will not give complete details; rather, we will refer the reader to the relevant references.

Notation. In this section, $i = \sqrt{-1}$ and $\overline{c}$ is the complex conjugate of $c \in \mathbb{C}$.

As explained in Section 1.3.2 and used after Theorem 2.8, we will start with the fact that if n is an integer, then

$$\int_0^1 e^{2\pi i n x}\, dx = \begin{cases} 1 & \text{if } n = 0\,; \\ 0 & \text{otherwise.} \end{cases}$$

Hence, letting $\mathcal{P}$ be the set of primes, we define

$$g(x) = \sum_{p \in \mathcal{P}} e^{2\pi i p x}$$

so that

$$\int_0^1 g^2(x) g(-2x)\, dx = \sum_{\substack{p_1, p_2, p_3 \in \mathcal{P} \\ p_1 + p_2 = 2p_3}} 1.$$

In the summation, note that $p + p = 2p$ is counted along with the desired 3 distinct primes in arithmetic progression (which are double counted as both p_1, p_3, p_2 and p_2, p_3, p_1). To link the integral to primes in (non-trivial) arithmetic progression, we restrict to the interval $[1, n]$ and define

$$A(n) = \sum_{\substack{p_1, p_2, p_3 \in \mathcal{P} \cap [1,n] \\ p_1 + p_2 = 2p_3}} 1 = |\mathcal{P} \cap [1, n]| + 2 \cdot S(n),$$

where $S(n)$ is the number of 3-term arithmetic progressions of primes in $[1, n]$.

If we can show that $A(n)$ is asymptotically larger than $|\mathcal{P} \cap [1, n]|$ then we can conclude we have infinitely many 3-term arithmetic progressions in $\mathcal{P}$. First, is this reasonable? Since, we are dealing with primes, we will be using the most famous of results on primes, the Prime Number Theorem:

Theorem 5.30 (Prime Number Theorem). *Let $\mathcal{P}_n = \mathcal{P} \cap [1, n]$. Then*

$$|\mathcal{P}_n| = \frac{n}{\log n}(1 + o(1)) \qquad \textit{and} \qquad \sum_{p \in \mathcal{P}_n} \log p = n(1 + o(1)).$$

From the Prime Number Theorem we know that about $\frac{1}{\log n}$ of the integers in $[1, n]$, concentrated along the odd integers, are prime. So, the "chance" that an odd integer is prime is about $\frac{2}{\log n}$, implying that the approximate "chance" that three terms in an arithmetic progression are all prime is $\left(\frac{2}{\log n}\right)^3$. There are about $\frac{n^2}{16}$ arithmetic progressions of length three consisting of only odd integers, so we expect around $\frac{n^2}{2(\log n)^3}$ arithmetic progressions of length three to consist of only prime numbers. Of course, this assumes a uniform distribution of primes, which we know is false. However, this approximation is surprisingly close: Grosswald [92] proved that

$$\frac{n^2}{3(\log n)^3}$$

is the correct order of magnitude.

To get an understanding of how we approach our integral

$$\int_0^1 g^2(x)g(-2x)\,dx,$$

we should start with the basic building block

$$g(x) = \sum_{p\in\mathcal{P}} e^{2\pi ipx}.$$

It turns out that it is better to use a weighted counting function:

$$f_n(x) = \sum_{p\in\mathcal{P}_n} (\log p)\, e^{2\pi ipx},$$

but the idea is still the same: we want to show that the contribution of 3-term arithmetic progressions in

$$\mathcal{P}_n = \mathcal{P} \cap [1, n]$$

is more than the contribution of $\mathcal{P}_n$.

Since our goal is to show that $S(n)$ is asymptotically larger than $|\mathcal{P}_n|$, we need a way to link this weighted counting function to $S(n)$. Using the trivial bound

$$f_n(x) \leq (\log n) \sum_{p\in\mathcal{P}_n} e^{2\pi ipx},$$

we see that

$$\frac{1}{\log^3 n}\int_0^1 f_n^2(x)f_n(-2x)\,dx \leq A(n) = |\mathcal{P}_n| + 2S(n).$$

Hence, if we can show that

$$\frac{1}{(\log n)^3}\int_0^1 f_n^2(x)f_n(-2x)\,dx$$

is asymptotically larger than $|\mathcal{P}_n| \approx \frac{n}{\log n}$, then we can conclude that $S(n) \to \infty$, which is our goal.

To get a feel for how our integral behaves, we start by considering its building block $f_n(x)$. Consider the graph of the real part of

$$f_{53}(x) = \sum_{p \in \mathcal{P}_{53}} (\log p)\, e^{2\pi i p x},$$

given in Figure 5.1.

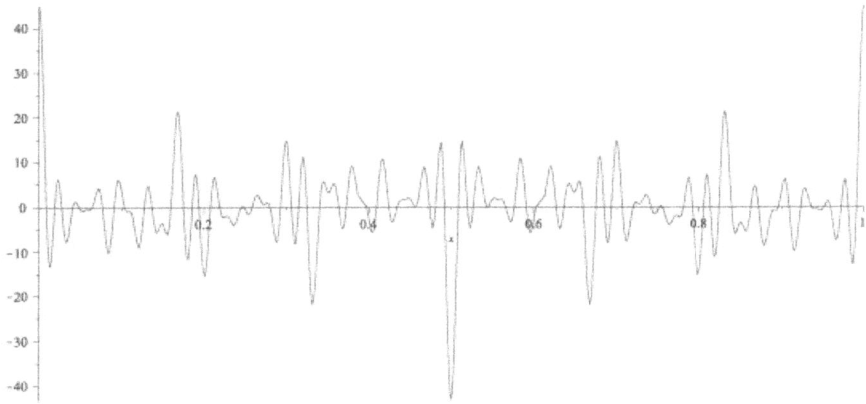

FIGURE 5.1
Plot of the real part of $\sum_{p \in \mathcal{P}_{53}} (\log p)\, e^{2\pi i p x}$

The spikes at $x = 0, 1$, and $\frac{1}{2}$ in Figure 5.1 can be explained readily. At $x = 0$ and $x = 1$ we have $\sum_{p \in \mathcal{P}_n} \log p$. By the Prime Number Theorem, this is $n(1 + o(1))$. At $x = \frac{1}{2}$, aside from $p = 2$, all primes are odd so $e^{2\pi i \frac{p}{2}} = -1$. We again see via the Prime Number Theorem that this gives $-n(1 + o(1))$.

Considering other parts of the graph in Figure 5.1, notice that if we are close to some fraction $\frac{a}{q}$ with $(a, q) = 1$ and q small, then we have, in general, larger values (in absolute value) than when we are not close to such a fraction. This is due to the cyclic nature of $\sum_{p \in \mathcal{P}_n} e^{2\pi i p x}$ for such fractions; we don't expect as much cancellation as with more "random" behavior on the unit circle.

Having the idea that "most" of the weight $f_n(x)$ is around $x = \frac{a}{q}$ with $a, q \in \mathbb{Z}^+$ relatively prime and q "small," we define, for $\epsilon > 0$, the intervals

$$I(a, q; \epsilon) = \left(\frac{a}{q} - \epsilon, \frac{a}{q} + \epsilon \right).$$

By restricting q to "small" values and taking ϵ suitably small, we can require $I(a, q; \epsilon) \cap I(a', q'; \epsilon) = \emptyset$ for $(a, q) \neq (a', q')$. In this situation, we call each $I(a, q; \epsilon)$ a *major arc*.

Let

$$\mathfrak{M} = \bigsqcup_{\substack{1\le a<q\\(a,q)=1\\q\ \text{small}}} I(a,q;\epsilon) \qquad \text{and} \qquad \mathfrak{m} = [0,1)\setminus\mathcal{M}$$

represent the *major arcs* and *minor arcs*, respectively.

The moniker major refers to the weight of the function and not $|\mathfrak{M}|$ since

$$|\mathfrak{M}| = 2\epsilon\sum_{q<Q}\phi(q) < \epsilon\, Q^2,$$

and taking $\epsilon = \frac{Q}{n}$ where Q^3 is small compared to n (e.g., $Q = \log n$), shows that $|\mathfrak{M}| \to 0$ as $n \to \infty$.

Our approach is to split up our integral over these arcs:

$$\int_0^1 f_n^2(x)f_n(-2x)\,dx = \int_{\mathfrak{M}} f_n^2(x)f_n(-2x)\,dx + \int_{\mathfrak{m}} f_n^2(x)f_n(-2x)\,dx$$

with the idea that the first integral on the right-hand side dominates the second integral on the right-hand side.

5.4.1.1 Minor Arcs

We start by estimating

$$\int_{\mathfrak{m}} f_n^2(x)f_n(-2x)\,dx.$$

We have

$$\begin{aligned}
\int_{\mathfrak{m}} f_n^2(x)f_n(-2x)\,dx &\le \sup_{x\in\mathfrak{m}}|f_n(x)|\int_{\mathfrak{m}}|f_n^2(x)|\,dx\\
&\le \sup_{x\in\mathfrak{m}}|f_n(x)|\int_0^1|f_n^2(x)|\,dx\\
&= \sup_{x\in\mathfrak{m}}|f_n(x)|\int_0^1 f_n(x)\overline{f_n(x)}\,dx\\
&= \sup_{x\in\mathfrak{m}}|f_n(x)|\int_0^1 f_n(x)f_n(-x)\,dx\\
&= \sup_{x\in\mathfrak{m}}|f_n(x)|\sum_{p\in\mathcal{P}_n}(\log p)^2\\
&\le \sup_{x\in\mathfrak{m}}|f_n(x)|(\log n)\sum_{p\in\mathcal{P}_n}\log p\\
&= \sup_{x\in\mathfrak{m}}|f_n(x)|(\log n)n(1+o(1)).
\end{aligned}$$

On the minor arcs, Vinagradov [210] has shown that, for any $D > 0$,

$$\sup_{x\in\mathfrak{m}}|f_n(x)| \le \frac{n}{(\log n)^D}(1+o(1)).$$

This is not an easy bound and the reader is referred to [206] for a detailed development. Using this bound, we can conclude that (using $D = 2$)

$$\int_{\mathfrak{m}} f_n^2(x) f_n(-2x)\,dx \leq \frac{n^2}{\log n}(1 + o(1)). \tag{5.6}$$

5.4.1.2 Major Arcs

We move on to an estimate of

$$\int_{\mathfrak{M}} f_n^2(x) f_n(-2x)\,dx.$$

We will appeal to the following three commonly used results, stated without proof. Details for Hardy and Wright's Möbius formula can be found in [102]. The Siegel-Walfisz Theorem as stated in Theorem 5.32 is weaker than the original (as done by Walfisz [212]). The partial sum formula can be found in most any appropriate analysis book.

Theorem 5.31 (Hardy and Wright's Möbius Formula). *Let $q, m \in \mathbb{Z}^+$ with $(q, m) = 1$. Then*

$$\sum_{\substack{1 \leq a \leq q \\ (a,q)=1}} e^{2\pi i \frac{a}{q} m} = \mu(q) = \begin{cases} 1 & \text{if } q = 1\,; \\ (-1)^k & \text{if } q \text{ is the product of } k \text{ distinct primes}; \\ 0 & \text{otherwise.} \end{cases}$$

Theorem 5.32 (Siegel-Walfisz Theorem). *Let $B > 0$. Let $n, q \in \mathbb{Z}^+$ with $q < (\log n)^B$. Let a be an integer with $(a, q) = 1$. Then*

$$\sum_{\substack{p \in \mathcal{P}_n \\ p \equiv a \pmod{q}}} \log p = \frac{n}{\phi(q)} + O\left(\frac{n}{(\log n)^C}\right)$$

for any $C > 0$, where ϕ is Euler's totient function. Consequently,

$$\sum_{\substack{p \in \mathcal{P}_n \\ p \equiv a \pmod{q}}} \log p \geq \frac{n}{(\log n)^B}(1 + o(1)).$$

Theorem 5.33 (Partial Sums). *Let $h : \mathbb{R} \to \mathbb{C}$ be continuously differentiable, let $\{a_j\}_{j=1}^{\infty} \subseteq \mathbb{C}$, and let $n \in \mathbb{Z}^+$. Then*

$$\sum_{j \leq n} a_j h(j) = h(n) \sum_{j \leq n} a_j - \int_1^n \left(h'(t) \sum_{j \leq t} a_j \right) dt.$$

In what follows, we will be using $1 + o(1)$ notation and be loose with our analysis as the goal is to give the reader an idea of how the circle method is used. The actual error terms are important for a rigorous development.

On each major arc $I(a, q; \epsilon)$ we have $f_n(x) \approx f_n(\frac{a}{q})$ so we will analyze $f_n(\frac{a}{q})$.

Given q we have

$$\begin{aligned} f_n\left(\frac{a}{q}\right) &= \sum_{p \in \mathcal{P}_n} (\log p) e^{2\pi i \frac{a}{q} p} \\ &= \sum_{r=1}^{q} \sum_{\substack{p \in \mathcal{P}_n \\ p \equiv r \ (\text{mod } q)}} (\log p) e^{2\pi i \frac{a}{q} p} \\ &= \sum_{r=1}^{q} e^{2\pi i \frac{a}{q} r} \sum_{\substack{p \in \mathcal{P}_n \\ p \equiv r \ (\text{mod } q)}} \log p. \end{aligned}$$

We impose the condition $(r, q) = 1$ on the inner sum to obtain

$$f_n\left(\frac{a}{q}\right) \geq \sum_{r=1}^{q} e^{2\pi i \frac{a}{q} r} \sum_{\substack{p \in \mathcal{P}_n \\ p \equiv r \ (\text{mod } q) \\ (r,q)=1}} \log p.$$

With this restriction, we can apply the Siegel-Walfisz Theorem (Theorem 5.32) for $q < (\log n)^B$ so that we have

$$f_n\left(\frac{a}{q}\right) \geq \left(\sum_{\substack{1 \leq r \leq q \\ (r,q)=1}} e^{2\pi i \frac{r}{q} a} \right) \frac{n}{\phi(q)} (1 + o(1)).$$

Applying Hardy and Wright's Möbius formula from Theorem 5.31, we obtain

$$f_n\left(\frac{a}{q}\right) \geq \frac{\mu(q)}{\phi(q)} n(1 + o(1)) \tag{5.7}$$

for $q < (\log n)^B$ for any given $B > 0$.

Using a similar development, but with a bit more care (and a more general version of Hardy and Wright's Möbius formula), we obtain

$$f_n\left(\frac{-2a}{q}\right) \geq \begin{cases} \dfrac{\mu(q)}{\phi(q)} n(1 + o(1)) & \text{if } q \text{ is odd}; \\ \dfrac{\mu(q/2)}{\phi(q/2)} n(1 + o(1)) & \text{if } q \text{ is even}. \end{cases} \tag{5.8}$$

Going back to the major arcs, let $\alpha \in I(a, q; \epsilon)$ so that $f_n(\alpha) \approx f_n(\frac{a}{q})$. We want to bound $|f_n(\alpha) - f_n(\frac{a}{q})|$. We have $f_n(\alpha) = f_n(\frac{a}{q} + \beta)$ for some $\beta \in (-\epsilon, \epsilon)$. Hence,

$$f_n(\alpha) = f_n\left(\frac{a}{q} + \beta\right) = \sum_{p \in \mathcal{P}_n} (\log p) e^{2\pi i (\frac{a}{q} + \beta) p} = \sum_{p \in \mathcal{P}_n} (\log p) e^{2\pi i \beta p} e^{2\pi i \frac{a}{q} p}$$

Using the Partial Sums formula from Theorem 5.33 and the triangle inequality, it turns out that

$$\left| f_n(\alpha) - \frac{\mu(q)}{\phi(q)} \sum_{j=1}^{n} e^{2\pi i \beta j} \right| = o(n)$$

and

$$\left| f_n(-2\alpha) - \begin{cases} \dfrac{\mu(q)}{\phi(q)} \displaystyle\sum_{j=1}^{n} e^{2\pi i(-2\beta)j} & \text{if } q \text{ is odd}\,; \\ \dfrac{\mu(q/2)}{\phi(q/2)} \displaystyle\sum_{j=1}^{n} e^{2\pi i(-2\beta)j} & \text{if } q \text{ is even}\,; \end{cases} \right| = o(n).$$

Note that the summation index is no longer over the primes in the above equations.

From here, the details become quite involved. We will give the highlights and encourage the reader to consult more detailed resources (e.g., [92, 206]).

With quite a bit of work, we obtain (roughly)

$$f_n^2(\alpha) f_n(-2a) \gtrsim \frac{(-1)^{q+1}\mu(q)}{(\phi(q))^3} \left(\sum_{j=1}^{n} e^{2\pi i \beta j} \right)^2 \left(\sum_{j=1}^{n} e^{2\pi i(-2\beta)j} \right) (1 + o(1)),$$

where we have used $(\mu(q))^3 = \mu(q)$ (by definition).

Next, we see that

$$\left(\sum_{j=1}^{n} e^{2\pi i \beta j} \right)^2 \left(\sum_{j=1}^{n} e^{2\pi i(-2\beta)j} \right)$$

is the generating function for the number of 3-term arithmetic progressions in $[1, n]$. Hence, we know that

$$\int_0^1 \left(\sum_{j=1}^{n} e^{2\pi i \beta j} \right)^2 \left(\sum_{j=1}^{n} e^{2\pi i(-2\beta)j} \right) d\beta = \frac{n^2}{4}(1 + o(1)).$$

The idea behind using the major arcs is that this is where most of the arithmetic progressions are being counted.

First, it is easy to see that integrating over $0 \leq \beta \leq 1$ is the same as integrating over $-\frac{1}{2} \leq \beta \leq \frac{1}{2}$. We will show that

$$\int_{-\epsilon}^{\epsilon} \left(\sum_{j=1}^{n} e^{2\pi i \beta j} \right)^2 \left(\sum_{j=1}^{n} e^{2\pi i(-2\beta)j} \right) d\beta = \frac{n^2}{4}(1 + o(1)) \qquad (5.9)$$

for $\epsilon = \frac{\log n}{2n}$.

Let

$$E(\beta) = \sum_{j=1}^{n} e^{2\pi i \beta j}.$$

Using

$$\left| \sum_{j=1}^{n} e^{2\pi i \beta j} \right| = \left| e^{2\pi i \beta} \right| \cdot \left| \frac{1 - e^{2\pi i \beta n}}{1 - e^{2\pi i \beta}} \right| \leq \frac{1}{|\beta|}$$

and

$$\left| \sum_{j=1}^{n} e^{2\pi i (-2\beta) j} \right| \leq n$$

we get

$$\left| \int_{\epsilon}^{\frac{1}{2}} (E(\beta))^2 \, E(-2\beta) \, d\beta \right| \leq n \int_{\epsilon}^{\frac{1}{2}} \beta^{-2} \, d\beta \leq \frac{n}{\epsilon}.$$

Hence,

$$\begin{aligned} \int_{-\epsilon}^{\epsilon} (E(\beta))^2 \, E(-2\beta) \, d\beta &= \int_{-\frac{1}{2}}^{\frac{1}{2}} (E(\beta))^2 \, E(-2\beta) \, d\beta - 2 \int_{\epsilon}^{\frac{1}{2}} (E(\beta))^2 \, E(-2\beta) \, d\beta \\ &\geq \frac{n^2}{4}(1 + o(1)) - \frac{2n}{\epsilon}. \end{aligned}$$

Provided we take

$$\epsilon \gg \frac{1}{n},$$

Equation (5.9) holds.

Summing over all major arcs and noting that $\epsilon = \frac{\log n}{2n} \gg \frac{1}{n}$ is sufficient to produce disjoint major arcs, we have

$$\begin{aligned} \int_{\mathfrak{M}} f_n^2(\alpha) f(-2\alpha)) \, d\alpha &= \sum_{q \leq (\log n)^B} \sum_{\substack{1 \leq a \leq q \\ (a,q)=1}} \int_{I(a,q;\epsilon)} f_n^2(\alpha) f_n(-2\alpha) \, d\alpha \\ &= \frac{n^2}{4} \left(\sum_{q \leq (\log n)^B} \sum_{\substack{1 \leq a \leq q \\ (a,q)=1}} \frac{(-1)^{q+1} \mu(q)}{(\phi(q))^3} (1 + o(1)) \right) \\ &= \frac{n^2}{4} \left(\sum_{q \leq (\log n)^B} \frac{(-1)^{q+1} \mu(q)}{(\phi(q))^2} \right) (1 + o(1)). \end{aligned}$$

It turns out that the last series above is $O(1)$; more precisely, it is roughly $\frac{4}{3}(1 + o(1))$ (where the choice of B is buried in the details).

Putting together the major and minor arc estimates, and recalling that we used a weighted generating function (weighted by $\log p$), we get

$$\begin{aligned}\int_0^1 \sum_{p\in\mathcal{P}_n} (\log p)e^{2\pi ipx}\,dx &= \int_{\mathfrak{M}} \sum_{p\in\mathcal{P}_n} (\log p)e^{2\pi ipx}\,dx + \int_{\mathfrak{m}} \sum_{p\in\mathcal{P}_n} (\log p)e^{2\pi ipx}\,dx \\ &\gtrsim \frac{n^2}{3}(1+o(1)) - \frac{n^2}{\log n}(1+o(1)) \\ &= \frac{n^2}{3}(1+o(1)).\end{aligned}$$

Accounting for the weighting, we conclude that the primes contain at least

$$\frac{n^2}{3(\log n)^3}(1+o(1))$$

arithmetic progressions of length three.

There are easier ways to prove that the primes contain infinitely many 3-term arithmetic progressions: see Estermann [72] and Chudakov [41, 42] for details (which both stem from Vinogradov [211]). The aim of the above presentation was to give the reader an idea of the beginnings of what has become a very fruitful area of research. As a powerful example, by creating deep extensions of the above techniques, Green and Tao [90] proved that the primes contain arbitrarily long arithmetic progressions.

5.5 Exercises

5.1 Prove that $d(x, y)$ as defined in Section 5.1.1 is a metric.

5.2 Show that the shift operator used in the proof of van der Waerden's Theorem in Section 5.1.1 is a homeomorphism; that is, show that it is a bijection and that both T and T^{-1} map open sets to open sets.

5.3 In the proof of van der Waerden's Theorem given in Section 5.1.1, we only need to require $d(T^{id}(y), y) < 1$ near the end of the proof. What conclusion can you draw if we consider $d(T^{id}(y), y) < \frac{1}{2^\ell}$? If we take $\ell > r$, what can you conclude?

5.4 Let S^1 be the unit circle and let $T(\theta)$ be rotation by θ. Let θ be irrational. Prove that (S^1, T) is a minimal topological dynamical system. With the obvious measure on S^1, prove that the system is an ergodic measure-preserving dynamical system.

5.5 Let $\epsilon, \theta > 0$. Prove that $\{n \in \mathbb{Z}^+ : |e^{2\pi i n\theta} - 1| < \epsilon\}$ is syndetic. Let T be rotation of the unit circle by $2\pi\theta$. Deduce that for any subset of the unit circle A with $|A| > \epsilon$ the set $\{n \in \mathbb{Z}^+ : T^n(A) \cap A \neq \emptyset\}$ contains arbitrarily long arithmetic progressions (without appealing to Furstenberg's Multiple Recurrence Theorem (Theorem 5.20)).

Remark. We can define a 2-coloring χ of $\mathbb{Z}^+$ based on this by letting $\chi(i) = 1$ if $T^i(A) \cap A \neq \emptyset$ and $\chi(i) = 0$ otherwise. Viewing this as an infinite binary string, if $2\pi\theta$ is irrational, the result is a Sturmian word.

5.6 Prove Lemma 5.19 by showing that the set of probability measures on $\{0,1\}^{\mathbb{Z}}$ is closed. Start by considering $P \in \{0,1\}^{\mathbb{Z}}$ that is an accumulation point of a sequence of probability measures. Show that $P(\emptyset) = 0$, $P(X) = 1$, and $P(A \sqcup B) = P(A) + P(B)$ for any disjoint sets $A, B \subseteq X$ so that P is a probability measure.

5.7 Prove or disprove: For any $\epsilon > 0$, if $S \subseteq \mathbb{Z}^+$ with $\overline{d}(S) > \epsilon$, then S contains $FS(T)$ for some infinite $T \subseteq \mathbb{Z}^+$.

5.8 Prove that $(\beta\mathbb{Z}^+, \oplus)$ is a semigroup; that is, show that $p \oplus q \in \beta\mathbb{Z}^+$ and that $p \oplus (q \oplus r) = (p \oplus q) \oplus r$.

5.9 Let p_n be the principle ultrafilter corresponding to $n \in \mathbb{Z}^+$. Show that $p_n \oplus p_m = p_{n+m}$, thereby showing how addition in $\mathbb{Z}^+$ extends to addition in $\beta\mathbb{Z}^+$.

5.10 Let $n \in \mathbb{Z}^+$. An unnamed mathematician has deemed certain subsets of $[1,n]$ *bad*. You have determined that if B_1 and B_2 are both bad sets, then $B_1 \cup B_2$ is also bad. Call a set $G \subseteq [1,n]$ *good* if $[1,n] \setminus G$ is bad. Let $\mathcal{G}$ be the set of all good subsets. Given that $\emptyset$ is bad, show that $\mathcal{G}$ is an ultrafilter on $[1,n]$ and deduce that there exists a unique $m \in \mathbb{Z}^+$ such that $\{m\}$ is good. Deduce that every ultrafilter on a finite set is a principle ultrafilter. What can be said if we replace $[1,n]$ with $\mathbb{Z}^+$?

5.11 This application is from [121]. Consider an arbitrary set of voters V voting for candidates from a finite set of candidates C. Each voter has an ordered preference list of candidates. Define the outcome to be the determination of an overall ordered preference of candidates (based on all voters' preferences).

Consider the following voting conditions:

(i) If all voters in V give the same ordered preference, then that ordering is the outcome.

(ii) The ranking of candidate a before candidate b in the outcome depends only on their relative orderings in the voters' preferences, and not on how any other candidates are ranked.

Define

$$\mathcal{F} = \Big\{ X \in \wp(V) : \text{for all } a, b \in C, \text{if } a \text{ is ranked before } b \text{ by all } x \in X, \text{then } a \text{ is ranked before } b \text{ in the outcome} \Big\}.$$

Prove that $\mathcal{F}$ is an ultrafilter.

5.12 Let $\mathcal{A}$ be the set of subsets of $\mathbb{Z}^+$ that contain arbitrarily long arithmetic progressions. Show that $\mathcal{A}$ is not a filter. Next, change part (iii) of Definition 5.22 to read: (iii) if $A \cup B \in p$ then $A \in p$ or $B \in p$. If a family satisfies (i) and (ii) of Definition 5.22 along with this revised part (iii), we say the family is a *superfilter*. Show that $\mathcal{A}$ is a superfilter. Prove that any ultrafilter is a superfilter.

5.13 Let $\mathcal{P}$ be the set of primes. First, prove that

$$\frac{6}{\pi^2} = \prod_{p \in \mathcal{P}} \left(1 - \frac{1}{p^2}\right).$$

Use this to show that the set of square-free integers contains arbitrarily long arithmetic progressions.

Hint:

$$\prod_{p \in \mathcal{P}} (1 - 1/p^2) = \prod_{p \in \mathcal{P}} \frac{1}{(1 - 1/p^2)^{-1}} = \prod_{p \in \mathcal{P}} \frac{1}{\sum_{k=0}^{\infty} (1/p^2)^k} = \frac{1}{\sum_{n=0}^{\infty} (1/n^2)}$$

5.14 The following result is due to Varnavides [207]: For $k \geq 3$ and $\epsilon > 0$, let n be sufficiently large and let $A \subseteq [1, n]$ with $|A| > \epsilon n$. Then there exists a constant $c = c(k, \epsilon) > 0$ such that A contains at least cn^2 arithmetic progressions of length k. Prove this via the following steps.

(a) Denote by $N(k, \epsilon)$ the integer from Szemerédi's Theorem (Theorem 2.70) so that for all $n > N(k, \epsilon)$, any set with more than ϵn elements contains a k-term arithmetic progression. Let $N \gg N(k, \epsilon/2)$ and let $K = N(k, \epsilon/2)$. Let $AP(d)$ be the set of K-term arithmetic progressions in $[1, N]$ with common difference d and let

$$AP = \bigcup_{d=1}^{\frac{N}{K-1}} AP(d).$$

Let $S = \{s \in AP : |A \cap s| \geq \frac{\epsilon}{2} K\}$. Show that $|S| \geq \frac{\epsilon^2}{32K} N^2$ by verifying that, for $d < \frac{\epsilon}{8K} N$, both

$$\sum_{s \in AP(d)} |A \cap s| \geq K |A \cap [Kd, N - Kd]| \geq K(|A| - 2Kd) \geq \frac{3\epsilon K}{4} N$$

and

$$\sum_{s\in AP(d)} |A\cap s| \;=\; \sum_{s\in AP(d)\cap S} |A\cap s| + \sum_{s\in AP(d)\setminus S} |A\cap s|$$
$$\le K|AP(d)\cap S| + \frac{\epsilon K}{2}N$$

hold. Conclude that for each $d \in \left[1, \frac{\epsilon}{8K}N\right)$ we have

$$|S\cap AP(d)| \ge \frac{\epsilon}{4}N$$

to prove the stated bound.

(b) Show that S contains the desired number of k-term arithmetic progressions as follows. Recall that $K = N(k, \epsilon/2)$ and that if $s \in S$ then $|A\cap s| \ge \frac{\epsilon}{2}K$. Use Szemerédi's Theorem to deduce the existence of $b, c \in \mathbb{Z}^+$ such that

$$\{b, b+c, \ldots, b+(k-1)c\} \subseteq A\cap s.$$

Prove that the number of $s \in AP$ associated with the same b and c is at most $\frac{1}{2}K^2$. Deduce the final result.

6

The Probabilistic Method

Even the odd one out is in with a shout.

–John Douglas, Francis Reader, & Paul Livingston

There seems to be no easy way to transition from the depth of the material in the previous chapter to this chapter. Like induction, using the probabilistic method can sometimes seems as if we are getting something for practically nothing.

The main idea is simple. Consider all r-colorings of a given n-element structure (e.g., K_n or $[1, n]$). Assume we are guaranteed some particular monochromatic substructure. We can conclude that the probability of a randomly chosen r-coloring admitting a monochromatic substructure is 1. If we can prove that for certain (smaller) values of n, this probability is strictly less than 1, then we know that there exists an r-coloring that does not admit a monochromatic substructure. Hence, we can deduce lower bounds for Ramsey-type numbers.

In this chapter we will investigate the probabilistic method's usefulness as it relates to Ramsey theory. For a thorough understanding of the method, the reader is urged to read Alon and Spencer's authoritative book [5] on the subject. For graph colorings specifically, see also Molloy and Reed's book [146].

6.1 Lower Bounds on Ramsey, van der Waerden, and Hales-Jewett Numbers

We start with a very basic, but effective use of the probabilistic method on the Ramsey numbers. Consider a random r-coloring of the edges of K_n so that each of the $r^{\binom{n}{2}}$ possible colorings is equally likely. For any ordering of the $\binom{n}{k}$ possible K_k subgraphs, let X_i be the random variable indicating whether or not the ith subgraph is monochromatic:

$$X_i = \begin{cases} 1 & \text{if the } i^{\text{th}}\ K_k \text{ is monochromatic;} \\ 0 & \text{otherwise.} \end{cases}$$

Consider

$$\mathbb{P}\left(\bigcup_{i=1}^{\binom{n}{k}} \{X_i = 1\}\right).$$

This is the probability that at least one X_i equals 1, i.e., that at least one of the K_k subgraphs is monochromatic. If this probability is strictly less than 1, then there must exist an r-coloring of the edges of K_n with no monochromatic K_k subgraph. Hence, we would have $R(k;r) > n$.

We next turn to the Principle of Inclusion-Exclusion (see Theorem 1.3). Since this is typically first encountered with events (as opposed to random variables), let $A_1, A_2, \ldots, A_m$ be events. Then

$$\mathbb{P}\left(\bigcup_{i=1}^{m} A_i\right) = \sum_{i=1}^{m} \sum_{\substack{|I| \subseteq [1,m] \\ |I|=i}} (-1)^{ji+1}\, \mathbb{P}\left(\bigcap_{j \in I} A_j\right). \tag{6.1}$$

For our purposes, $m = \binom{n}{k}$ and A_i is the event that $X_i = 1$.

Unfortunately, for Ramsey-type numbers this is practically impossible to calculate (see [218] for more details). We do know, applying one of the Bonferroni inequalities (Theorem 1.4), that

$$\mathbb{P}\left(\bigcup_{i=1}^{\binom{n}{k}} A_i\right) \leq \sum_{i=1}^{\binom{n}{k}} \mathbb{P}(A_i), \tag{6.2}$$

and this is something we can handle (with ease).

Remark. The reader may wonder at this point why we introduced random variables, only to swiftly change to events. This is because introductory probability classes typically couch the Principle of Inclusion-Exclusion in terms of events. However, we will be using random variables later and most other resources frame statements in terms of random variables.

On the right-hand side of Inequality (6.2) we easily have

$$\mathbb{P}(A_i) = \frac{r}{r^{\binom{k}{2}}}$$

(r choices for the color and each of the $\binom{k}{2}$ edges of the K_k have that color, independently, with probability $\frac{1}{r}$) so that

$$\sum_{i=1}^{\binom{n}{k}} \mathbb{P}(A_i) = \frac{\binom{n}{k}}{r^{\binom{k}{2}-1}}.$$

We now set this quantity to be strictly less than 1 and solve for n. Using Stirling's approximation, we have

$$\binom{n}{k} \approx \frac{1}{\sqrt{2\pi k}}\left(\frac{ne}{k}\right)^k.$$

Hence, we want the maximal n such that

$$\frac{1}{\sqrt{2\pi k}}\left(\frac{ne}{k}\right)^k < r^{\binom{k}{2}-1}.$$

We have

$$n < \frac{\left(\frac{2\pi k}{r}\right)^{\frac{1}{k}} k}{e} r^{\frac{k}{2}-\frac{1}{2}}.$$

For k sufficiently large, $\left(\frac{2\pi k}{r}\right)^{\frac{1}{k}} \approx 1$, so we can conclude that

$$R(k;r) \geq \frac{k}{e} \cdot r^{\frac{k-1}{2}}(1+o(1)). \tag{6.3}$$

This is not the best-known bound; we will encounter the current (as of this writing) best-known lower bound in Section 6.4. However, as a gauge for the effectiveness of the basic probabilistic method used above, the bound in Inequality (6.3) was discovered by Erdős [63] in 1947 and had not been fundamentally changed until very recently (a modestly improved bound in Section 6.4 moves $\sqrt{r}$ from the denominator to the numerator, and was done in 1975).

Next, we will apply the probabilistic method to the van der Waerden numbers $w(k;r)$. The number of k-term arithmetic progressions in $[1,n]$ is

$$\frac{n^2}{2(k-1)}(1+o(1)).$$

Order these in any way and let X_i be the indicator random variable for when the ith arithmetic progression is monochromatic. As in the $R(k;r)$ example above, we have (suppressing the error term in the number of arithmetic progressions)

$$\mathbb{P}\left(\bigcup_{i=1}^{\frac{n^2}{2(k-1)}} \{X_i = 1\}\right) \leq \sum_{i=1}^{\frac{n^2}{2(k-1)}} \mathbb{P}(X_i = 1).$$

Under a random r-coloring of $[1,n]$ we have $P(X_i = 1) = \frac{r}{r^k}$ for every i. Hence,

$$\mathbb{P}\left(\bigcup_{i=1}^{\frac{n^2}{2(k-1)}} \{X_i = 1\}\right) \leq \frac{n^2}{2(k-1)r^{k-1}}(1+o(1)).$$

Setting this strictly less than 1, we obtain

$$n < \sqrt{2(k-1)}r^{\frac{k-1}{2}}(1-o(1)),$$

so we can conclude that

$$w(k;r) \geq \sqrt{2(k-1)}r^{\frac{k-1}{2}}(1-o(1))$$

for sufficiently large k.

It turns out that, although this is a non-trivial bound, it is much weaker than the best-known bound (see Theorem 2.4). We attempt to explain this weaker bound in a different context in Section 6.3.

As our final example, we apply the probabilistic method to more difficult-to-calculate numbers: the Hales-Jewett numbers. The only non-trivial number known is $HJ(3;2) = 4$, calculated by Hindman and Tressler [112] (trivially, $HJ(2;r) = r$ and $HJ(k,1) = 1$).

Recall that $HJ(k;r) = n$ means that n is the minimal integer such that any r-coloring of the length n words over the alphabet $\{1, 2, \ldots, k\}$ admits a monochromatic variable word. There are $(k+1)^n - k^n$ variable words of length n (recall that a variable word must include at least one occurrence of the variable).

Now, for a variable word $w(x)$ to be monochromatic, we require $w(1), w(2), \ldots, w(k)$ to all have the same color. If the words are randomly colored, then the probability of this occurring is $\frac{r}{r^k}$.

For $1 \leq i \leq (k+1)^n - k^n$, let X_i be the indicator random variable equaling 1 if and only if the ith variable word is monochromatic. Proceeding as we did in the previous two examples, we have

$$\mathbb{P}\left(\bigcup_{i=1}^{(k+1)^n-k^n} \{X_i = 1\}\right) \leq \sum_{i=1}^{(k+1)^n-k^n} \mathbb{P}(X_i = 1) = \frac{1}{r^{k-1}}((k+1)^n - k^n).$$

Setting

$$\frac{1}{r^{k-1}}((k+1)^n - k^n) < 1$$

we want to find n so that

$$(k+1)^n - k^n < r^{k-1}. \tag{6.4}$$

If $k \leq r-1$, we may take $n = k-1$ and satisfy this inequality. Hence, we have $HJ(k;r) \geq k$ for $k \leq r-1$. However, we typically want to let k grow while r is fixed, so we now assume that $k \geq r$ and we assume k is large enough so that $e^{\frac{n}{k}}$ is a good approximation of $\left(1+\frac{1}{k}\right)^n$. Using this approximation in Inequality (6.4) we want

$$k^n\left(e^{\frac{n}{k}} - 1\right) < r^{k-1}.$$

We will not attempt to optimize this; rather, we choose simplicity and opt to consider

$$k^n e^{\frac{n}{k}} \leq r^{k-1}.$$

From here it is easy to conclude that

$$HJ(k;r) \geq \frac{k(k-1)\ln r}{k\ln k + 1}$$

for $k \geq r$ sufficiently large.

This is not a good bound and is presented only as an application of the basic probabilistic method that fails rather spectacularly. As we have seen in Corollary 2.53, we know that $HJ(k;r)$ is exponential for fixed r. In fact, the probabilistic method bound is worse than the original bound of $HJ(k;2) > k$ given in [96].

The decreasing effectiveness of the basic probability method of the three examples in this section is explained by the increasing degree of dependence of the random variables in each example.

6.2 Turán's Theorem

As promised in Section 3.5, we now offer another proof of the bound on the number of edges in Turán's Theorem. The application of the probabilistic method in this situation is more involved than those found in the last section.

For reference, here is the statement we will prove:

(Partial) Turán's Theorem. *Let $G = (V, E)$ be a graph on n vertices with no K_k subgraph. Then*

$$|E| \leq \frac{k-2}{2(k-1)}n^2.$$

We will be using the linearity of expectation. We present this result as the next lemma.

Lemma 6.1 (Linearity of Expectation). *Let $X_1, X_2, \ldots, X_n$ be random variables. Then*

$$\mathbb{E}\left(\sum_{i=1}^{n} X_i\right) = \sum_{i=1}^{n} \mathbb{E}(X_i).$$

Proof. It suffices to prove this for $n = 2$ as the induction is obvious. We will also assume that the random variables are discrete since that is what we will be using (and the continuous case follows readily). Let X and Y have joint probability mass function $f(x,y) = \mathbb{P}(X = x, Y = y)$ with marginal mass functions $f_X(x)$ and $f_Y(y)$. Then

$$\begin{aligned}\mathbb{E}(X+Y) = \sum_x \sum_y (x+y) f(x,y) &= \sum_x \sum_y x f(x,y) + \sum_x \sum_y y f(x,y) \\ &= \sum_x x \sum_y f(x,y) + \sum_y y \sum_x f(x,y) \\ &= \sum_x x f_X(x) + \sum_y y f_Y(y) \\ &= \mathbb{E}(X) + \mathbb{E}(Y),\end{aligned}$$

as required. □

Note that we do not require independence to apply Lemma 6.1. This is why it will be useful for us: all of the indicator random variables we have used are dependent on some (but not all) of the other random variables. It is also useful to recall that for indicator random variables,

$$\mathbb{E}(X) = \mathbb{P}(X = 1)$$

by definition of expectation.

The argument we present below is due to Y. Caro and V. Wei (according to [5, p. 100]).

Proof of Turán's Theorem. Using the colors red and blue, we consider the statement of the theorem as G has no blue K_k subgraph, and prove the bound as a bound on the number of blue edges. We use a random greedy algorithm to "construct" a blue complete subgraph on vertex set S.

Let $V = \{1, 2, \ldots, n\}$ and take a random permutation π of V. Initialize $S = \emptyset$. For i from 1 to n do the following:

if $\pi(i)$ has no red neighbor in S, then $S = S \cup \{\pi(i)\}$.

Clearly, the subgraph on S is a blue complete subgraph. We want to bound the expected size of S. To this end let

$$X_v = \begin{cases} 1 & \text{if } v \in S; \\ 0 & \text{otherwise.} \end{cases}$$

Using Lemma 6.1 along with

$$\mathbb{E}(X_v) = \mathbb{P}(X_v = 1)$$

since X_v is an indicator random variable, we have

$$\mathbb{E}(|S|) = \mathbb{E}\left(\sum_{v \in V} X_v\right) = \sum_{v \in V} \mathbb{E}(X_v) = \sum_{v \in V} \mathbb{P}(X_v = 1).$$

Let $\deg_r(v)$ be the number of vertices connected to v with a red edge. Then $\mathbb{P}(X_v = 1)$ means that v precedes all of its red neighbors in π. Since there are $(\deg_r(v)+1)!$ ways to order v along with its red neighbors, of which $\deg_r(v)!$ have v first, we see that

$$\mathbb{P}(X_v = 1) = \frac{1}{\deg_r(v)+1}.$$

Hence, we have

$$\mathbb{E}(|S|) = \sum_{v \in V} \frac{1}{\deg_r(v)+1}.$$

We now appeal to Jensen's Inequality, which states that for any convex function f and any $x_1, x_2, \ldots, x_m$ in its domain, we have

$$\frac{\sum_{i=1}^m f(x_i)}{m} \geq f\left(\frac{\sum_{i=1}^m x_i}{m}\right).$$

Applying this with $f(x) = \frac{1}{x+1}$ (which is convex for $x \geq 0$; i.e., concave-up by the second derivative test), $m = |V|$, and $x_i = \deg_r(v_i)$, we get

$$\frac{1}{|V|} \sum_{v \in V} \frac{1}{\deg_r(v)} \geq \frac{1}{\overline{x}+1},$$

where

$$\overline{x} = \frac{1}{|V|} \sum_{v \in V} \deg_r(v),$$

i.e., the average number of red neighbors in G.

Let E_r and E_b be the number or red, respectively, blue, edges in G. Then the average number of red neighbors in G is

$$\frac{2E_r}{|V|}$$

since each edge accounts for two neighbors (one on each end of the edge). Hence, we have

$$\sum_{v \in V} \frac{1}{\deg_r(v)} \geq \frac{|V|}{\overline{x}+1} = \frac{|V|}{\frac{2E_r}{|V|}+1}.$$

By assumption, any blue complete subgraph has at most $k-1$ vertices so we have $\mathbb{E}(|S|) \leq k-1$. Thus,

$$\frac{|V|}{\frac{2E_r}{|V|}+1} \leq k-1,$$

which gives us

$$E_r \geq \frac{n^2}{2(k-1)} - \frac{n}{2}.$$

Since $E_r + E_b = \binom{n}{2}$ we get

$$E_b \leq \binom{n}{2} - \left(\frac{n^2}{2(k-1)} - \frac{n}{2}\right) = \frac{k-2}{2(k-1)}n^2,$$

as required. □

6.3 Almost-surely van der Waerden and Ramsey Numbers

Consider the following probabilistic take on van der Waerden numbers. We know that with $n = w(k;r)$, *all* r-colorings of $[1,n]$ admit a monochromatic k-term arithmetic progression. What happens if we relax this to "almost all?" Formally, let $X_k(n)$ be the number of monochromatic k-term arithmetic progressions in an r-coloring of $[1,n]$ and consider the following definition.

Definition 6.2. We say that $w(k;r) \leq h(k)$ *almost-surely*, and write $w(k;r) \overset{\text{a.s.}}{\leq} h(k)$, if

$$\lim_{k\to\infty} \mathbb{P}(X_k(h(k)) > 0) = 1.$$

We say that $w(k;r) > h(k)$ *almost-surely*, and write $w(k;r) \overset{\text{a.s.}}{>} h(k)$, if

$$\lim_{k\to\infty} \mathbb{P}(X_k(h(k)) = 0) = 1.$$

Of course, this definition can be applied to any Ramsey-type number. We will only apply it to the van der Waerden and Ramsey numbers here. Our first goal in this section is to prove the following result, which gives the order of these almost-surely van der Waerden numbers. The lower bound is due to Vijay [208], while the upper bound can be found in [169].

Theorem 6.3. *Let $f(k)$ and $g(k)$ tend to 0 and ∞, respectively, arbitrarily slowly. Then*

$$\sqrt{k}\, r^{\frac{k}{2}} f(k) \overset{\text{a.s.}}{<} w(k;r) \overset{\text{a.s.}}{\leq} \sqrt{k}\, r^{\frac{k}{2}} g(k),$$

in other words, $w(k;r) \overset{\text{a.s.}}{\approx} \sqrt{k}\, r^{\frac{k}{2}}$.

Before delving into the proof, notice that this result is essentially the bound given by the probabilistic method in Section 6.1. After the proof of Theorem 6.3, we show a markedly different situation for the almost-surely Ramsey numbers.

Proof of Theorem 6.3. We start with the lower bound, which follows smoothly from the linearity of expectation and the definition of expectation. Let

$$n = \sqrt{k}\, r^{\frac{k}{2}} f(k).$$

For $1 \leq i \leq \frac{n^2}{2(k-1)}$ (suppressing the error term), let X_i equal 1 if the ith arithmetic progression of length k is monochromatic under a given r-coloring, and equal 0 otherwise. Then the expected number of monochromatic k-term arithmetic progressions is

$$\mathbb{E}\left(\sum_{i=1}^{\frac{n^2}{2(k-1)}} X_i\right) = \sum_{i=1}^{\frac{n^2}{2(k-1)}} \mathbb{E}(X_i) = \sum_{i=1}^{\frac{n^2}{2(k-1)}} \mathbb{P}(X_i = 1)$$

$$= \frac{n^2}{2(k-1)r^{k-1}}$$

$$= \frac{rk}{2(k-1)} f^2(k).$$

By definition of expectation, letting

$$X = \sum_{i=0}^{\frac{n^2}{2(k-1)}} X_i,$$

we have

$$\mathbb{E}(X) = \sum_{i=0}^{\frac{n^2}{2(k-1)}} i\,\mathbb{P}(X = i) \geq \sum_{i=0}^{\frac{n^2}{2(k-1)}} \mathbb{P}(X = i) - \mathbb{P}(X = 0)$$

$$= 1 - \mathbb{P}(X = 0).$$

Hence, for all $k \in \mathbb{Z}^+$, we have

$$\frac{rk}{2(k-1)} f^2(k) \geq 1 - \mathbb{P}(X = 0)$$

so that

$$\mathbb{P}(X = 0) \geq 1 - \frac{rk}{2(k-1)} f^2(k) \underset{k\to\infty}{\longrightarrow} 1,$$

which proves that

$$w(k;r) \overset{\text{a.s.}}{>} n = \sqrt{k}\, r^{\frac{k}{2}} f(k).$$

The upper bound is more involved. The idea is to construct a subset of all k-term arithmetic progressions with limited dependence while having a positive proportion of all k-term arithmetic progressions. We then show that under random colorings, the probability that one of the arithmetic progressions from this subset is monochromatic tends to 1.

The limited dependence between elements of this subset will allow us to use effectively two terms of the Principle of Inclusion-Exclusion formula (Equation (6.1)).

The construction of such a subset is done by removing arithmetic progressions that are "close" to each other. To this end, let $AP(n)$ be all k-term arithmetic progressions in $[1, n]$ and denote by $A_d(n)$ those arithmetic progressions in $AP(n)$ with common difference d.

From each $A_d(n)$, remove every arithmetic progression with initial term in

$$\bigsqcup_{j=1}^{\left\lceil \frac{n}{3d} - \frac{k-1}{3} \right\rceil} [(3j-2)d + 1, 3jd].$$

Let $B_d(n)$ be the set of elements of $A_d(n)$ that are not removed and define

$$B(n) = \bigcup_{d=1}^{\left\lfloor \frac{n-1}{k-1} \right\rfloor} B_d(n).$$

Notice that, by construction, any two intersecting k-term arithmetic progressions in any $B_d(n)$ have initial terms that are more than $2d$ apart.

We now state, without proof, a technical result needed to prove the upper bound. For the proof, the reader is referred to [169].

`Fact.` Let $A_1, A_2 \in B(n)$ with common differences d_1 and d_2, respectively. The following hold:

(i) $\frac{|AP(n)|}{3} \leq |B(n)| \leq \frac{|AP(n)|}{2}$;

(ii) $|A_1 \cap A_2| \leq k - 3$;

(iii) $|A_1 \cap A_2| > \left\lceil \frac{k}{2} \right\rceil$ only if $d_1 = d_2$;

(iv) $\left\lceil \frac{k}{3} \right\rceil \leq |A_1 \cap A_2| \leq \left\lceil \frac{k}{2} \right\rceil$ only if $\frac{d_1}{d_2} \in \left\{\frac{1}{3}, \frac{1}{2}, \frac{2}{3}, 1, \frac{3}{2}, 2, 3\right\}$.

We now have what we need to prove the upper bound. Let

$$n = \sqrt{k}\, r^{\frac{k}{2}} g(k)$$

and partition $[1, n]$ into disjoint intervals of length

$$s = \left\lceil \frac{n}{g^{\frac{4}{3}}(k)} \right\rceil$$

(where the last interval may be shorter, but isn't needed).

We now consider only arithmetic progressions in $B(n)$.

Randomly r-color $[1, n]$ and consider a particular interval I of length s. Without loss of generality, we may assume that $I = [1, s]$.

Let F_a be the event that arithmetic progression $a \in B(s) = B(n) \cap I$ is monochromatic. Using a Bonferroni inequality (Theorem 1.4) on the Principle of Inclusion-Exclusion, we define the probability p and bound it as follows:

$$p = \mathbb{P}\left(\bigcup_{a \in B(s)} F_a \right) \geq \sum_{a \in B(s)} \mathbb{P}(F_a) - \sum_{b \in B(s)} \sum_{\substack{a \in B(s) \\ a_1 \geq b_1}} \mathbb{P}(F_a \cap F_b)$$

$$\geq \frac{s^2}{6(k-1)r^{k-1}} - \sum_{b \in B(s)} \sum_{\substack{a \in B(s) \\ a_1 \geq b_1}} \mathbb{P}(F_a \cap F_b), \qquad (6.5)$$

where we use the lower bound in Fact (i) in the second step, and a_1 and b_1 are the first terms of a and b, respectively.

We now bound the double sum. With $b \in B(s)$ fixed and having common difference d, consider the following partitioning:

$$S_b = \{a \in B_d(s) : a \cap b \neq \emptyset\};$$

$$T_b = \{a \in B_{d_2}(s) : \lceil \tfrac{k}{3} \rceil \leq |a \cap b| \leq \lceil \tfrac{k}{2} \rceil \text{ and } \tfrac{d}{d_2} \in \{\tfrac{1}{3}, \tfrac{1}{2}, \tfrac{2}{3}, 1, \tfrac{3}{2}, 2, 3\}\};$$

$$Q_b = \{a \in B(s) : a \cap b \neq \emptyset \text{ and } a \notin S_b \cup T_b\};$$

$$R_b = \{a \in B(s) : a \cap b = \emptyset\}.$$

Using (i)-(iv) of our Fact, we obtain the following, for k sufficiently large:

$$\sum_{a \in S_b} \mathbb{P}(F_a \cap F_b) \leq \sum_{i=2}^{k-1} \frac{1}{r^{k+i}} = \frac{1}{r^{k+1}} - \frac{1}{r^{2k-1}} < \frac{1}{r^{k+1}};$$

$$\sum_{a \in T_b} \mathbb{P}(F_a \cap F_b) \leq \frac{7\left(\frac{k}{2} - \frac{k}{3} + 1\right)}{r^{k+\frac{k}{2}}} \leq \frac{1}{3r^{k+1}};$$

$$\sum_{a \in Q_b} \mathbb{P}(F_a \cap F_b) \leq \frac{3sk}{r^{k+\frac{2k}{3}}} \leq \frac{1}{3r^{k+1}};$$

$$\sum_{a \in R_b} \mathbb{P}(F_a \cap F_b) \leq \frac{s^2}{4(k-1)r^{2k-2}} \leq \frac{1}{3r^{k+1}}.$$

The first inequality holds since there is exactly one arithmetic progression that intersects b in exactly $k-1-i$ places (this follows from Fact (iii) and them having the same common difference with $a_1 \geq b_1$). The second inequality holds by Fact (iv). The third bound uses the lower bound of Fact (iv) along with the fact that less than $3sk$ arithmetic progressions of k-terms can intersect a given k-term arithmetic progression (see Section 6.4 for more information on this). The last inequality uses the upper bound from Fact (i) along with the independence of the two events.

Putting this all together, we can conclude that for a fixed $b \in B(s)$, we have

$$\sum_{\substack{a \in B(s) \\ a_1 \geq b_1}} \mathbb{P}(F_a \cap F_b) \leq \frac{2}{r^{k+1}} \leq \frac{1}{r^k}.$$

Hence,

$$\sum_{b \in B(s)} \sum_{\substack{a \in B(s) \\ a_1 \geq b_1}} \mathbb{P}(F_a \cap F_b) \leq \frac{|B(s)|}{r^k} \leq \frac{s^2}{4(k-1)r^k}.$$

Going back to p (see Inequality (6.5)), we have

$$p \geq \frac{s^2}{6(k-1)r^{k-1}} - \frac{s^2}{4(k-1)r^k} = \frac{(2r-3)s^2}{12(k-1)r^k}.$$

This means that each interval of length s has probability

$$1 - p \leq 1 - \frac{(2r-3)s^2}{12(k-1)r^k}$$

of having no monochromatic k-term arithmetic progression.

Thus, the probability that none of the $\lfloor \frac{n}{s} \rfloor$ intervals of length s have a monochromatic k-term arithmetic progression is

$$(1-p)^{\lfloor \frac{n}{s} \rfloor} \leq (1-p)^{g^{\frac{4}{3}}(k)-1} \leq \left(1 - \frac{(2r-3)s^2}{12(k-1)r^k}\right)^{g^{\frac{4}{3}}(k)-1}.$$

For k sufficiently large, we obtain

$$\begin{aligned}
\left(1 - \tfrac{(2r-3)s^2}{12(k-1)r^k}\right)^{g^{\frac{4}{3}}(k)-1} &\leq \exp\left(-\tfrac{(g^{\frac{4}{3}}(k)-1)(2r-3)s^2}{12(k-1)r^k}\right) \\
&\leq \exp\left(-\tfrac{(2r-3)n^2}{g^{\frac{5}{3}}(k)12(k-1)r^k}\right) \\
&\leq \exp\left(-\tfrac{(2r-3)k}{12(k-1)} g^{\frac{1}{3}}(k)\right).
\end{aligned}$$

Since

$$\lim_{k \to \infty} \exp\left(-\frac{(2r-3)k}{12(k-1)} g^{\frac{1}{3}}(k)\right) = 0$$

we see that the probability of a random r-coloring of $[1, n]$ admitting a monochromatic k-term arithmetic progressions tends to 1, as required for the upper bound. □

As promised, we next consider the Ramsey numbers through an "almost-surely" lens. A lower bound is given in Exercise 6.11. Unfortunately, determining a good upper bound via a method similar to that used on the van der Waerden numbers seems unlikely. For the 2-color Ramsey number $n = R(k,k)$, we are searching for copies of K_k, denoted $G_1, G_2, \dots, G_b$, such that G_i and G_j share a minimal number of edges for $i \neq j$ but the union of all edges of the G_i consists of a positive proportion of all edges of K_n. We easily have, if all the G_i share no edges, that the probability of no G_i being monochromatic is

$$\left(1 - \frac{1}{2^{\binom{k}{2}-1}}\right)^b. \tag{6.6}$$

Our goal for an upper bound is to show that this tends to 0. Hence, we want to maximize $b = b(n,k)$.

To maximize b, consider $(2,k,n)$-Steiner systems. Assuming the existence of such a system, we have

$$b = \frac{\binom{n}{2}}{\binom{k}{2}} \approx \left(\frac{n}{k}\right)^2.$$

To have the expression in (6.6) tend to 0 we need $b = 2^{\binom{k}{2}} g(k)$ where $g(k) \to \infty$ (at any rate). Hence, we require

$$\left(\frac{n}{k}\right)^2 \approx 2^{\binom{k}{2}} g(k),$$

which gives

$$n \approx k\, 2^{\frac{k^2}{4}} \sqrt{g(k)}.$$

However, this is worse than the (non-probabilistic) upper bound $R(k,k) < \frac{4^{k-1}}{\sqrt{k-1}}$ (see Inequality (3.1) in Section 3.1).

It seems a more powerful probabilistic tool is needed. We visit one in the next section.

6.4 Lovász Local Lemma

A tool that has been quite successful for improving bounds for Ramsey-type numbers is the Lovász Local Lemma. It first appeared in a paper by Erdős and Lovász [67]. The idea is as follows. If events $\{A_i\}_i$ are mutually independent then we know that

$$\mathbb{P}\left(\bigcap_i A_i^c\right) = \prod_i (1 - \mathbb{P}(A_i)).$$

Can we find a similar result for events that have limited mutual dependency? For example, if we consider the K_k subgraphs of K_n with n large, then with event A_i being that the ith subgraph is monochromatic, we see that many (but not all) of the events are independent since many pairs of subgraphs do not share edges.

We have used

$$\mathbb{P}\left(\bigcap_i A_i^c\right) \geq 1 - \sum_i \mathbb{P}(A_i)$$

in previous sections of this chapter, so our goal is to provide a better bound than this. All we need to deduce to find lower bounds is

$$\mathbb{P}\left(\bigcap_i A_i^c\right) > 0$$

when A_i is the event that the ith structure is monochromatic in a random coloring.

Before stating the Lovász Local Lemma, we require the following definition.

Definition 6.4 (Dependency graph). Let $A_1, A_2, \ldots, A_m$ be events. Let $G = (V, E)$ be a graph on m vertices where $V = \{A_i\}_{i=1}^n$ and where $\{A_i, A_j\} \in E$ if and only if A_i and A_j are dependent. We call G a *dependency graph.*

Lemma 6.5 (Lovász Local Lemma). *Let $A_1, A_2, \ldots, A_m$ be events with dependency graph $G = (V, E)$. If there exist $x_1, x_2, \ldots, x_m \in [0, 1)$ such that*

$$\mathbb{P}(A_i) \leq x_i \prod_{\{A_i, A_j\} \in E} (1 - x_j)$$

for all $i \in \{1, 2, \ldots, m\}$, then

$$\mathbb{P}\left(\bigcap_{i=1}^m A_i^c\right) \geq \prod_{i=1}^m (1 - x_i) > 0.$$

As an immediate consequence, we obtain the symmetric form of the lemma.

Corollary 6.6 (Symmetric Lovász Local Lemma). *Let $A_1, A_2, \ldots, A_m$ be events with dependency graph $G = (V, E)$. If $\mathbb{P}(A_i) \leq p$ for $1 \leq i \leq m$, and $ep(d + 1) \leq 1$, where d is the maximal vertex degree of G, then*

$$\mathbb{P}\left(\bigcap_{i=1}^m A_i^c\right) > 0.$$

Proof. Take

$$x_i = \frac{1}{d + 1}$$

for $1 \leq i \leq m$ and note that

$$\mathbb{P}(A_i) \leq p \leq \frac{1}{d+1} \cdot \frac{1}{e} \leq \frac{1}{d+1}\left(1 - \frac{1}{d+1}\right)^d$$
$$\leq \frac{1}{d+1} \prod_{\{A_i, A_j\} \in E} \left(1 - \frac{1}{d+1}\right).$$

Applying Lemma 6.5 completes the proof. □

The proof of the Lovász Local Lemma essentially boils down to induction and a clever use of conditional probability. The proof below is adapted from the original proof found in [67].

Proof of Lemma 6.5. We induct on m, with $m = 1$ being trivial. It suffices to prove that

$$\mathbb{P}\left(A_1 \middle| \bigcap_{i=2}^{m} A_i^c\right) \leq x_1 \tag{6.7}$$

since then

$$\mathbb{P}\left(A_1^c \middle| \bigcap_{i=2}^{m} A_i^c\right) \geq 1 - x_1,$$

so that, by the definition of conditional probability, we have

$$\mathbb{P}\left(A_1^c \middle| \bigcap_{i=2}^{m} A_i^c\right) = \frac{\mathbb{P}\left(\bigcap_{i=1}^{m} A_i^c\right)}{\mathbb{P}\left(\bigcap_{i=2}^{m} A_i^c\right)} \geq 1 - x_1,$$

i.e.,

$$\mathbb{P}\left(\bigcap_{i=1}^{m} A_i^c\right) \geq (1 - x_1)\,\mathbb{P}\left(\bigcap_{i=2}^{m} A_i^c\right).$$

By the inductive assumption, the result follows.

To show Inequality (6.7), we (again) induct on m. The base case $m = 1$ is given, so we assume the result holds for $m - 1$ and will show Inequality (6.7) holds as shown.

We will use the following conditional probability result:

$$\mathbb{P}\left(\bigcap_{i=1}^{k} B_i \middle| C\right) = \prod_{i=1}^{k} \mathbb{P}\left(B_i \middle| \bigcap_{j=i+1}^{k} (B_j \cap C)\right). \tag{6.8}$$

After reordering, if necessary, let $A_2, A_3, \ldots, A_q$ be all events connected to

A_1 in G (we have $q \geq 2$, else there is nothing to show). Then

$$\mathbb{P}\left(A_1 \,\middle|\, \bigcap_{i=2}^{m} A_i^c\right) = \frac{\mathbb{P}\left(A_1 \cap \bigcap_{i=2}^{q} A_i^c \,\middle|\, \bigcap_{\ell=q+1}^{m} A_\ell^c\right)}{\mathbb{P}\left(\bigcap_{i=2}^{q} A_i^c \,\middle|\, \bigcap_{\ell=q+1}^{m} A_\ell^c\right)}.$$

Using a basic inequality on the numerator and applying Equation (6.8) to the denominator, we obtain

$$\mathbb{P}\left(A_1 \,\middle|\, \bigcap_{i=2}^{m} A_i^c\right) \leq \frac{\mathbb{P}\left(A_1 \,\middle|\, \bigcap_{\ell=q+1}^{m} A_\ell^c\right)}{\prod_{i=2}^{q} \mathbb{P}\left(A_i^c \,\middle|\, \bigcap_{\ell=i+1}^{m} A_\ell^c\right)}.$$

We now note that each term in the product in the denominator has at most $m-2$ terms in the condition so that we can apply the inductive assumption to each term. In the numerator, A_1 and the conditional event are independent, so we have

$$\mathbb{P}\left(A_1 \,\middle|\, \bigcap_{i=2}^{m} A_i^c\right) \leq \frac{\mathbb{P}(A_1)}{\prod_{i=2}^{q}(1-x_i)}. \tag{6.9}$$

We now apply the theorem's hypothesis to the numerator, to obtain

$$\mathbb{P}\left(A_1 \,\middle|\, \bigcap_{i=2}^{m} A_i^c\right) \leq \frac{x_1 \prod_{i=2}^{q}(1-x_i)}{\prod_{i=2}^{q}(1-x_i)} = x_1,$$

as required. □

The above proof is an existential proof and offers no insight into determining the x_i values to use in the Lovász Local Lemma. However, algorithmic approaches started with Beck [12] and have culminated in a general result by Moser and Tardos [150].

We will now apply the Lovász Local Lemma (in its symmetric form) to $R(k;r)$ as Spencer [194] did to provide a lower bound that stood for over 40 years (the $r = 2$ case is still the best-known lower bound).

Theorem 6.7. *Let $k, r \in \mathbb{Z}^+$. We have the following bound for the Ramsey numbers:*

$$R(k;r) \geq \frac{1}{e}\, k\, r^{\frac{k+1}{2}}(1+o(1)).$$

Proof. Let p and d be as in Corollary 6.6. Randomly color the edges of K_n with r colors. For $1 \le i \le \binom{n}{k}$, let A_i be the event that the ith complete subgraph on k vertices in monochromatic. Then

$$\mathbb{P}(A_i) = r^{1-\binom{k}{2}}$$

for all i. Hence, we take

$$p = r^{1-\binom{k}{2}}.$$

Next, we bound d. Note that in order for A_i and A_j to be dependent, the associated K_k subgraphs must share at least 2 vertices. Hence, for a fixed A_i there are at most

$$\binom{k}{2}\binom{n-2}{k-2}$$

events connected to A_i in the dependency graph. Thus, we take

$$d = \binom{k}{2}\binom{n-2}{k-2}.$$

We now bound $ep(d+1)$:

$$ep(d+1) = e\,\frac{\binom{k}{2}\binom{n-2}{k-2}+1}{r^{\binom{k}{2}-1}} \lesssim \frac{erk^2\binom{n-2}{k-2}}{2r^{\binom{k}{2}}} \approx \frac{erk^2}{2}\,\frac{\frac{k^2}{n^2}\binom{n}{k}}{r^{\binom{k}{2}}}.$$

We next apply Stirling's formula to obtain

$$ep(d+1) \lesssim \frac{erk^4}{2n^2}\frac{1}{r^{\binom{k}{2}}}\left(\frac{ne}{k}\right)^k.$$

Setting this quantity to be at most 1, we isolate n:

$$n^{k-2} \le \frac{2}{erk^4}\left(\frac{k}{e}\right)^k r^{\binom{k}{2}},$$

so that

$$n \le \frac{k}{e} r^{\frac{1}{2}\frac{k(k-1)}{k-2}}(1+o(1)) = \frac{k}{e} r^{\frac{1}{2}\frac{k(k-2)+k}{k-2}}(1+o(1))$$

$$= \frac{1}{e}\,k\,r^{\frac{k+1}{2}}(1+o(1)),$$

which is the desired bound. □

Remark. Our bound on d in the above proof is close to $\binom{n}{k}$, the total number of events. Hence, we have a lot of dependency. Clearly, the smaller the d, the better the bound and hence we don't see much improvement over the bound derived in Section 6.1; the Lovász Local Lemma works better when there is little dependency.

As the reader may have inferred, Theorem 6.7 no longer provides the best-known lower bound on the multicolor Ramsey numbers (it does for $r = 2$, though). In a very recent preprint, Conlon and Ferber [50] provide a substantial improvement by representing edges as vectors over finite fields along with randomly coloring orthogonal vectors. Their result is stated next.

Theorem 6.8. *Let $k, r \in \mathbb{Z}^+$. Then*

$$\begin{aligned} R(k;3r) &> 2^{\frac{7r}{8}k(1+o(1))}; \\ R(k;3r+1) &> 2^{\frac{7r+3(\log 3-1)}{8}k(1+o(1))}; \\ R(k;3r+2) &> 2^{\frac{7r+4}{8}k(1+o(1))}. \end{aligned}$$

Recall that the basic probability method bounds the van der Waerden numbers by $w(k;r) \geq \sqrt{2(k-1)}r^{\frac{k-1}{2}}(1-o(1))$. The Lovász Local Lemma drastically improves this bound.

Theorem 6.9. *Let $k, r \in \mathbb{Z}^+$. Then*

$$w(k;r) > \frac{r^{k-1}-e}{(k+2)e}.$$

Proof. Let p and d be as in Corollary 6.6. Consider a random r-coloring of $[1,n]$. For $1 \leq i \leq \frac{n^2}{2(k-1)}$, denote by A_i the event that the ith arithmetic progression of length k is monochromatic. Clearly, we may take $p = r^{1-k}$.

We next show that in the dependency graph we have a maximal degree of at most $(k+2)n$. Fix an event A_i and let P be the associated k-term arithmetic progression. Any other event A_j is dependent on A_i only if the two associated k-term arithmetic progressions intersect. We want to bound the number of k-term arithmetic progressions that intersect P. Assume that integer m is common to both arithmetic progressions. There are k choices for which term m is in P. If Q is a k-term arithmetic progression that also contains m, then there are k choices for which term m is in Q, and less than $\frac{n}{k-1}$ possible common differences for Q. Hence, the number of possible A_j connected to a fixed A_i in the dependency graph is at most

$$\frac{k^2}{k-1}n \leq (k+2)n.$$

Applying Corollary 6.6, we set

$$ep(d+1) \leq er^{1-k}((k+2)n+1) \leq 1.$$

Solving for n, we obtain

$$n \leq \frac{r^{k-1}-e}{(k+2)e},$$

which yields the stated bound. $\square$

The bound in Theorem 6.9 stood for quite some time; however, the current best-known bound is cr^{k-1} (see Theorem 2.4).

We now investigate off-diagonal Ramsey-type numbers. We will restrict to two colors. The results should be good if we fix one variable and let the other be large. This will decrease dependency of events.

We will present the best-known lower bound for $w(k,\ell)$ with ℓ fixed. The bound is from [134] and uses Lemma 6.5. Unlike in the proofs using the symmetric form of the Lovász Local Lemma, the colors are not equally likely.

Theorem 6.10. *Let $k,\ell \in \mathbb{Z}^+$ with ℓ fixed. Then*

$$w(k,\ell) > \left(\frac{\ell-1}{\ell^2}\right)^{\ell}\left(\frac{k}{\log k}\right)^{\ell-1}$$

for k sufficiently large.

Proof. Let n be the lower bound given in the statement of the theorem. We will show that there exists a 2-coloring of $[1,n]$ with no red ℓ-term arithmetic progression and no blue k-term arithmetic progression. For each $i \in [1,n]$, color i red with probability

$$\frac{\ell}{\ell-1}\cdot\frac{\log kn}{k}$$

and blue otherwise. We have two types of events: Type A is the occurrence of a red ℓ-term arithmetic progression; Type B is the occurrence of a blue k-term arithmetic progression. For any Type A event we have

$$\mathbb{P}(A) = \left(\frac{\ell}{\ell-1}\right)^{\ell}\frac{(\log kn)^{\ell}}{k^{\ell}},$$

while for a Type B event we have

$$\mathbb{P}(B) = \left(1-\frac{\ell}{\ell-1}\frac{\log kn}{k}\right)^{k} \le e^{-\frac{\ell}{\ell-1}\log kn} = \frac{(\log k)^{\ell}}{k^{\ell+\ell/(\ell-1)}} = q$$

for k sufficiently large.

We now bound the maximal degree in the dependency graph, where clearly two vertices are connected if and only if their associated arithmetic progressions intersect. There are four types of intersections to consider. For a fixed Type A event, we let d_{AA} be the maximal number of dependent Type A events and let d_{AB} be the maximal number of dependent Type B events. For a fixed Type B event, we let d_{BA} be the maximal number of dependent Type A events and let d_{BB} be the maximal number of dependent Type B events.

As explained in the proof of Theorem 6.9, the bounds below follow:

$$d_{AA} \le \frac{\ell^2}{\ell - 1} n;$$

$$d_{AB} \le \frac{k\ell}{k - 1} n;$$

$$d_{BA} \le \frac{\ell k}{\ell - 1} n;$$

$$d_{BB} \le \frac{k^2}{k - 1} n.$$

Once we determine $a, b \in [0, 1)$ such that

$$\mathbb{P}(A) \le a(1 - a)^{d_{AA}}(1 - b)^{d_{AB}};$$
$$\mathbb{P}(B) \le b(1 - a)^{d_{BA}}(1 - b)^{d_{BB}}, \tag{6.10}$$

by Lemma 6.5, the result follows.

For convenience, let $\widehat{a} = a\,\mathbb{P}(A)$ and $\widehat{b} = b\,\mathbb{P}(B)$. The inequalities in (6.10) become

$$\log \widehat{a} \ge -d_{AA} \log(1 - \widehat{a}) - d_{AB} \log(1 - \widehat{b});$$

$$\log \widehat{b} \ge -d_{BA} \log(1 - \widehat{a}) - d_{BB} \log(1 - \widehat{b}).$$

Letting

$$\widehat{b} = \frac{1}{nkq} \qquad \text{and} \qquad \widehat{a} = \left(\widehat{b}\right)^{\frac{\ell}{k}}$$

and using a bit of algebra and asymptotic approximations shows that the needed inequalities are satisfied. For more details, see [134]. □

Via an analysis similar to that in the previous proof, Spencer [193] proved the following for off-diagonal Ramsey numbers (Spencer's proof was first chronologically). The proof is left to the reader as Exercise 6.9.

Theorem 6.11. *Let $k, \ell \in \mathbb{Z}^+$ with ℓ fixed. Then there exists a constant $c > 0$ such that*

$$R(k, \ell) \ge c\left(\frac{k}{\log k}\right)^{\frac{\ell+1}{2}}$$

for k sufficiently large.

Remark. The exponent of the denominator in Theorem 6.11 has been improved for $\ell \ge 5$ to $\frac{\ell+1}{2} - \frac{1}{\ell-2}$ in [26].

Analyzing the proof of Lemma 6.5 we see that we can generalize to a superset of the set of dependency graphs. To obtain Inequality (6.9) we used

$$\mathbb{P}\left(A_1 \,\middle|\, \bigcap_{\ell=q+1}^{m} A_\ell^c\right) = \mathbb{P}(A_1),$$

but all we need for the proof to hold is

$$\mathbb{P}\left(A_1 \,\middle|\, \bigcap_{\ell=q+1}^{m} A_\ell^c\right) \leq \mathbb{P}(A_1),$$

This leads to the following definition.

Definition 6.12 (c-negative dependency graph). Let $c \geq 1$. Let $A_1, A_2, \ldots, A_m$ be events. Let $G = (V, E)$ be a graph on m vertices where $V = \{A_i\}_{i=1}^m$. For each A_i, let $N(i)$ be the set of vertices connected to A_i along with A_i. If for every $S \subseteq V$ with $S \cap N(i) = \emptyset$ we have

$$\mathbb{P}\left(A_i \,\middle|\, \bigcap_{A_j \in S} A_j^c\right) \leq c\,\mathbb{P}(A_i)$$

for every A_i, then we say G is a *c-negative dependency graph.* For $c = 1$, we simply call G a *negative dependency graph.*

With a negative dependency graph we would have the events be that certain structures are monochromatic. We could loosely state that a negative dependency graph has the property that if we know a set of structures not connected to a given structure all fail to be monochromatic, then the probability that the given structure is monochromatic does not increase. For a c-negative dependency graph, we could state that the potential increase in probability is controlled.

Using this definition, we can state the generalization to supersets of dependency graphs. The $c = 1$ case is referred to as the Lopsided Lovász Local Lemma. The proof is very similar to that of Lemma 6.5 and is left to the reader as Exercise 6.13.

Lemma 6.13. *Let $c \geq 1$. Let $G = (V, E)$ be a c-negative dependency graph on events $A_1, A_2, \ldots, A_m$. If there exist $x_1, x_2, \ldots, x_m \in [0, \frac{1}{c})$ such that*

$$\mathbb{P}(A_i) \leq x_i \prod_{\{A_i, A_j\} \in E} (1 - c x_j)$$

for all $i \in \{1, 2, \ldots, m\}$, then

$$\mathbb{P}\left(\bigcap_{i=1}^{m} A_i^c\right) \geq \prod_{i=1}^{m}(1 - c x_i) > 0.$$

Thus far, Lemma 6.13 has not been successfully applied to improve upon the best-known bounds of any Ramsey-type number. The difficulty is in defining appropriate c-negative dependency graphs. For more information on negative dependency graphs and how they can be used, see [138] and [145].

6.5 Exercises

6.1 Let $X_1, X_2, \ldots, X_n$ be indicator random variables. Let $X = \sum_{i=1}^{n} X_i$. Prove that if $\mathbb{E}(X) < 1$ then $\mathbb{P}(X = 0) > 0$. Note that this is the basic probabilistic method in a nutshell.

6.2 Show that in a random 4-coloring of $[1, 40]$, using the colors $\{a, h, m, t\}$, the expected number of occurrences of the string $math$ occurring in arithmetic progression is more than 1.

6.3 Consider a graph on n vertices. Let $p = \frac{\frac{k-2}{2(k-1)}n^2}{\binom{n}{2}}$ and create an edge between any two vertices independently with probability p. Show that the expected number of edges is the maximum allowed by Turán's Theorem if we desire the graph to not contain any K_k subgraph. Under this edge-creation probability, what is the expected number of K_k subgraphs?

6.4 Consider an ℓ-uniform hypergraph G on n vertices with the property that every vertex is a member of exactly k edges (such a hypergraph is called *k-regular*). Let $k \leq 2^{\frac{\ell-1}{2}}$. Show that G is 2-colorable for ℓ sufficiently large.

6.5 Use the basic probability method to derive the following bound for the hypergraph Ramsey numbers:

$$R_\ell(k; r) \geq \frac{k}{e} \cdot r^{\frac{1}{\ell}\binom{k-1}{\ell-1}}(1 - o(1)).$$

What improvement, if any, does the Lovász Local Lemma offer?

6.6 Consider the 2-dimensional van der Waerden numbers (see Definition 3.33 and Corollary 3.35). Let $k, r \in \mathbb{Z}^+$ and denote these numbers by $w_2(k; r)$; that is, $w_2(k; r)$ is the minimal integer n such that any r-coloring of $[1, n] \times [1, n]$ admits a monochromatic 2-dimensional arithmetic progression of dimension length k. Show that

$$w_2(k; r) \geq \left(\frac{3k}{4}\right)^{\frac{1}{3}} r^{\frac{k^2-1}{3}}(1 - o(1)).$$

Hint:

The number of 2-dimensional arithmetic progressions of dimension length k in $[1, n]^2$ is

$$2\sum_{d=1}^{\frac{n-1}{k-1}} \sum_{a=1}^{n-(k-1)d} \sum_{b=1}^{a} 1 + O(n^2).$$

Use $\sum_{i=1}^{m} i^2 = \frac{m^3}{3}(1 + o(1))$.

6.7 Let $k, n \in \mathbb{Z}^+$ with $n \geq 2k$ and consider a family $\mathcal{F}$ of k-element subsets of $[1, n]$ with the property that for all $A, B \in \mathcal{F}$ we have $A \cap B \neq \emptyset$. The Erdős-Ko-Rado Theorem [66] states that

$$|\mathcal{F}| \leq \binom{n-1}{k-1}.$$

Prove this via probabilities by assuming that every k-element subset is equally likely to be in $\mathcal{F}$.

Hint:

Consider the n subsets $\{1, 2, \ldots, k\}, \{2, 3, \ldots, k+1\}, \ldots, \{n, 1, 2, \ldots, k-1\}$. What is the maximal number of subsets from $\mathcal{F}$ that can appear in these?

6.8 Consider the number $n(k; 2)$ from Theorem 2.55. Show that for some constant $c > 0$,

$$n(k; 2) > 2^{c\frac{k^2}{\log k}}$$

via the basic probabilistic method along with the following results.

(a) For $S \subseteq [1, n]$ with $|S| = k$, show that $|FS(S)| \geq \binom{k+1}{2}$.

(b) Show that at most $(kn)^{\log j} j^{2k}$ subsets $S \subseteq [1, n]$ with $|S| = k$ have $|FS(S)| \leq j$.

Hint:

See [69].

6.9 Let $k, \ell \in \mathbb{Z}^+$ with $\ell \geq 3$ fixed and k sufficiently large. Show that there exists a constant $c > 0$ such that

$$R(k, \ell) \geq c\left(\frac{k}{\log k}\right)^{\frac{\ell+1}{2}}.$$

Hint:

See [193] where only a sketch is provided.

6.10 Let $n, k \in \mathbb{Z}^+$ with k fixed. Show that

$$\lim_{n\to\infty} \binom{n}{k} p_n^k (1 - p_n)^{n-k} = e^{-\lambda}\frac{\lambda^k}{k!}$$

for any sequence of probabilities $\{p_n\}$ provided $\lambda = \lim_{n\to\infty} np_n$ is finite. (This is often referred to as the Law of Rare Events and explains why the Poisson distribution is a limiting distribution in many situations.)

6.11 Show that for the r-color Ramsey numbers we have

$$R(k;r) \overset{\text{a.s.}}{>} k\,r^{\frac{k}{2}} f(k),$$

where $f(k) \to 0$ arbitrarily slowly.

6.12 Let's attempt a framework to bound almost-surely off-diagonal van der Waerden numbers. We'll stick with 2 colors, say, red and blue. Let $\ell \in \mathbb{Z}^+$ be associated with red and $k \in \mathbb{Z}^+$ with blue. We can no longer assume that each integer in $[1,n]$ is equally likely to be red as it is to be blue. Let $\mathbb{P}(i \text{ is red}) = p = \frac{2^k}{2^\ell+2^k}$. For the probability space based on this p, show that, when $\ell \leq k$, we have

$$w(k,\ell) \overset{\text{a.s.}}{>} \sqrt{\ell}\left(\frac{2^\ell+2^k}{2^k}\right)^{\frac{\ell}{2}} f(k,\ell),$$

where $f(k,\ell) \to 0$ arbitrarily slowly. What bound do you obtain when we change the probability space to one based on $p = \frac{k}{\ell+k}$?

6.13 Prove Lemma 6.13.

7

Applications

I've sung one too many songs for a crowd that didn't want to hear.

–David Schelzel

Unbreakable, inevitable, assured, inescapable; that's how we started this book. We have now visited many properties that must persist under partitioning. This principle can be applied in effective ways, and should be heeded in others. We visit just a few of these situations in this chapter. For a more thorough investigation of Ramsey theory applications, see [173] and references therein.

7.1 Fermat's Last Theorem

Fermat's Last Theorem states that for $n \in \mathbb{Z}^+$ with $n \geq 3$, the equation $x^n + y^n = z^n$ has no solution in the positive integers. It was proved in 1995 by Wiles [214], but many attempts were made in the (roughly) 350 years between its conjecture and its proof. One obvious way to attempt solving this problem is to show that the modular equivalent has no solution for primes, i.e., that we cannot solve $x^n + y^n \equiv z^n \pmod{p}$ with $xyz \not\equiv 0 \pmod{p}$ for any prime p. Unfortunately for many amateur mathematicians, this has been shown to be a false statement. In 1901, Dickson [57] showed that for n an odd prime,

$$x^n + y^n \equiv z^n \pmod{p}$$

with $xyz \not\equiv 0 \pmod{p}$ has a solution for all sufficiently large primes p (he gives the bound $p \geq (n-1)^2(n-2)^2 + 6n - 2$). Fifteen years later, Schur [186] gave a proof for all positive integer exponents. We present two proofs of this result. The first is Schur's original argument, relying on abstract algebra; the second uses Fourier analysis as found in [140].

Theorem 7.1. *Let $n \in \mathbb{Z}^+$. There exists an integer $s(n)$ such that $x^n + y^n \equiv z^n \pmod{p}$ has a solution with $xyz \not\equiv 0 \pmod{p}$ for any prime $p > s(n)$.*

Proof 1. Consider the group $\mathbb{Z}_p^* = \mathbb{Z}_p \setminus \{0\}$ under multiplication. Let

$$T = \{i^n : i \in \mathbb{Z}_p^*\}$$

and note that T is a subgroup of $\mathbb{Z}_p^*$. Hence, for $r = (n, p-1) \le n$ (we know that T has $\frac{p-1}{r}$ elements and all cosets of T are the same size) there exist $a_1, a_2, \ldots, a_r \in \mathbb{Z}_p^*$ such that

$$\mathbb{Z}_p^* = \bigsqcup_{i=1}^{r} a_i T.$$

Hence, we have a partition of $[1, p-1]$. As such, we define the r-coloring $\chi : \mathbb{Z}_p^* \to \{1, 2, \ldots, r\}$ by

$$\chi(m) = i \text{ if and only if } m \in a_i T.$$

Let $s(r)$ be the r-color Schur number and ensure that $p - 1 \ge s(r)$. By Schur's Theorem (Theorem 2.15), we have $i, j, k \in \mathbb{Z}_p^*$ of the same color such that $i + j = k$. This means that

$$i = a_t x^n, \; j = a_t y^n, \text{ and } k = a_t z^n$$

for some $a_t, x, y, z \in \mathbb{Z}_p^*$. Hence,

$$a_t x^n + a_t y^n \equiv a_t z^n \pmod{p}.$$

As $\mathbb{Z}_p^*$ is a group, a_t^{-1} exists, so we may multiply both sides (on the left) by a_t^{-1} to prove the congruence. Since we do not have r in the statement of the theorem, the proof is finished by noting that $s(r) \le s(n)$ since $r \le n$ so that taking $p - 1 \ge s(n)$ suffices. □

Our second proof uses some Fourier analysis, so the reader may find reviewing Sections 1.3 and 2.5.1 helpful. We use $i = \sqrt{-1}$ and the complex conjugate notation $\overline{c}$ for this proof.

Proof 2. Let p be prime and, for $k, n \in \mathbb{Z}^+$ (we may assume $n \ge 2$ since $n = 1$ is obvious), define

$$S_k(n) = \sum_{j=0}^{p-1} e^{2\pi i \frac{kj^n}{p}}.$$

We start by obtaining a needed estimate (this is not the obvious starting point). Using $S_k = S_k(n)$ we have

$$\sum_{k \in \mathbb{Z}_p} |S_k|^2 = \sum_{k \in \mathbb{Z}_p} S_k \overline{S_k} = \sum_{k \in \mathbb{Z}_p} \sum_{x \in \mathbb{Z}_p} e^{2\pi i \frac{kx^n}{p}} \sum_{y \in \mathbb{Z}_p} e^{-2\pi i \frac{ky^n}{p}}$$

$$= \sum_{(x,y) \in \mathbb{Z}_p \times \mathbb{Z}_p} \sum_{k \in \mathbb{Z}_p} e^{2\pi i \frac{k(x^n - y^n)}{p}}.$$

Let N_p be the set of $(x, y) \in \mathbb{Z}_p \times \mathbb{Z}_p$ with $x^n - y^n \equiv 0 \pmod{p}$. Using

$$\frac{S_k(1)}{p} = \begin{cases} 1 & \text{if } k \equiv 0 \pmod{p}; \\ 0 & \text{otherwise,} \end{cases}$$

we conclude that

$$\sum_{k \in \mathbb{Z}_p} |S_k|^2 = p|N_p|.$$

We next bound $|N_p|$. We have

$$N_p = \{(0,0)\} \cup \{(x, ux) : x \neq 0, u^n \equiv 1 \pmod{p}\}.$$

The latter set has $p - 1$ choices for x and $(n, p-1) \leq n$ choices for u. Hence

$$|N_p| \leq 1 + (p-1)n < np,$$

so we conclude that

$$\sum_{k \in \mathbb{Z}_p} |S_k|^2 < p^2 n.$$

As in Proof 1, we let $\mathbb{Z}_p^* = \mathbb{Z}_p \setminus \{0\}$ be the group under multiplication and consider the cosets of $T = \{j^n : j \in \mathbb{Z}_p^*\}$. Then there exist $a_1, a_2, \ldots, a_r \in \mathbb{Z}_p^*$ such that

$$\mathbb{Z}_p^* = \bigsqcup_{j=1}^{r} a_i T,$$

where $r \leq n$. Since $a_j T \subseteq \mathbb{Z}_p^*$, for each $j \in \{1, 2, \ldots, r\}$ we have

$$\sum_{k \in a_j T} |S_k|^2 \leq \sum_{k \in \mathbb{Z}_p} |S_k|^2 < p^2 n.$$

Going back to the definition of $S_k(n)$, notice that $S_{cj^n}(n) = S_c(n)$ for any $j^n \in T$. This gives the following bound:

$$\sum_{k \in a_j T} |S_k|^2 = |T||S_{a_j}|^2 = \frac{p-1}{r}|S_{a_j}|^2 \geq \frac{p-1}{n}|S_{a_j}|^2.$$

Hence,

$$\frac{p-1}{n}|S_{a_j}|^2 \leq \sum_{k \in a_j T} |S_k|^2 < p^2 n$$

regardless of the value of $a_j \in \mathbb{Z}_p^*$. Thus, we conclude that, for any $k \in \mathbb{Z}_p^*$, we have

$$|S_k|^2 < \frac{p^2 n^2}{p-1} < 2pn^2.$$

Since we want to show that there exist $x, y, z \in \mathbb{Z}_p$ such that $x^n + y^n = z^n$, we find that the number of solutions is

$$F(n,p) = \sum_{(x,y,z)\in\mathbb{Z}_p^3} \frac{1}{p} \sum_{k=0}^{p-1} e^{2\pi i \frac{k(x^n+y^n-z^n)}{p}}.$$

Note that this sum includes solutions with $xyz \equiv 0 \pmod{p}$ so we must address this. First, we have

$$\begin{aligned} F(n,p) = \frac{1}{p}\sum_{k=0}^{p-1} S_k^2\,\overline{S_k} = \frac{1}{p}S_0^2\overline{S_0} + \frac{1}{p}\sum_{k=1}^{p-1} S_k^2\,\overline{S_k} &= p^2 + \frac{1}{p}\sum_{k=1}^{p-1} S_k^2\,\overline{S_k} \\ &\geq p^2 - \frac{1}{p}\sum_{k=1}^{p-1} |S_k|^3 \\ &> p^2 - \frac{1}{p}(p-1)(2pn^2)^{\frac{3}{2}} \\ &> p^2 - 2n^3 p^{\frac{3}{2}}. \end{aligned}$$

We now subtract solutions with at least one of x, y, and z divisible by p. We have three choices for which variable is divisible by p. For the other two variables, note that we want elements from N_p. Since $|N_p| < np$, we have at most $3np$ solutions with $xyz \equiv 0 \pmod{p}$. Hence, we have more than

$$p^2 - 2n^3 p^{\frac{3}{2}} - 3np$$

solutions with $xyz \not\equiv 0 \pmod{p}$. This quantity is strictly greater than 0 for

$$p \geq (n^3 + \sqrt{n^6 + 3n})^2,$$

which proves the theorem. □

While Proof 1 is the one presented by Schur [186], Proof 2 is more in-line with the proof provided by Dickson [57]. Also of note is that the bound on the prime p required to prove the result is smaller in the second proof. In general, Fourier series proofs provide better bounds than their Ramsey-theoretic counterparts.

7.2 Encoding Information

When sending messages across transmission lines (e.g., emails), the messages are encoded into a binary sequence. So, we may view message-sending as a stream of 1s and 0s, with each digit called a bit. Errors in transmission can crop up in many ways. We could have a random bit error (due to simple chance error), a burst error consisting of several interconnected simultaneous bit errors (due, for example, to some outside physical damage), crosstalk when other transmissions infiltrate the signal (self-infiltration is called an echo), or a noisy channel (due, for example, to electrical interference).

We investigate here the noisy channel. Consider a simple email with each character translated into a unique binary sequence. For example, the transmission of the character "]" is 01011100 and can be viewed as in Figure 7.1.

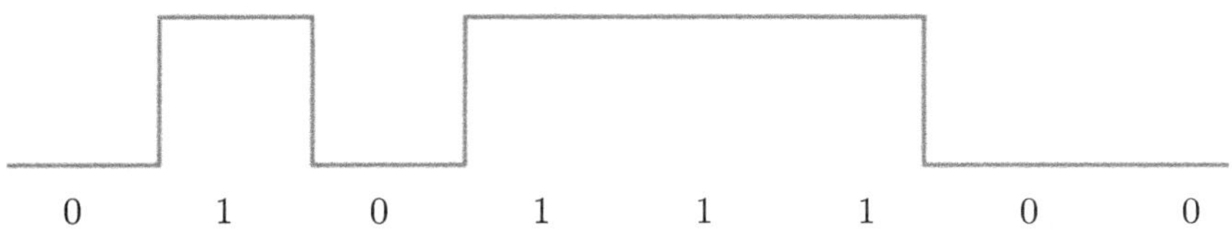

FIGURE 7.1
Pure signal of the character "]" with binary equivalent 01011100

However, because of a noisy channel, the received signal may look like Figure 7.2.

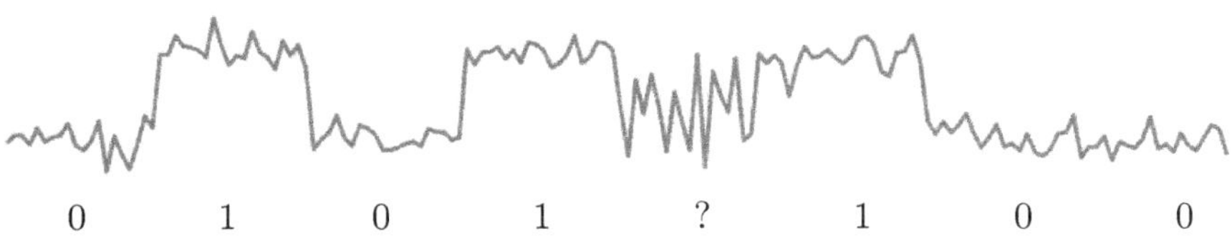

FIGURE 7.2
Receipt of 01011100 through a noisy channel

Hence, the received signal may be interpreted as 01010100, which is the binary equivalent of the character "U." So, on the receiving end we could confuse the character] with the letter U; however, it is highly unlikely (we'll assume impossible) for it to be confused with the symbol £, which is coded 10100011 (so that all bits are wrong). Such possible confusions lead naturally to a *confusion graph.* The set of vertices is the transmission alphabet and an edge between 2 letters exists if and only if confusion can arise, i.e., if and only if transmission through a noisy channel can cause a binary sequence associated with one vertex to be interpreted as the binary sequence associated with the other vertex.

To avoid confusion we may opt to choose a subset of the alphabet to be used so that no confusion may arise. In other words, we want to select an independent set of vertices in the confusion graph. We would then have an unambiguous code alphabet. For maximal flexibility, we want the largest such code alphabet, which corresponds to the largest independent set in a confusion graph.

Clearly, when we send messages we do not send one letter at a time. Hence, we need to understand how confusion graphs for longer strings can be created.

Definition 7.2 (Strong product of graphs). Let $G_1 = (V_1, E_1)$ and $G_2 = (V_2, E_2)$ be graphs. The *strong product of* G_1 *and* G_2 is $G_1 \boxtimes G_2 = (V_1 \times V_2, E)$ where (x_1, x_2) and (y_1, y_2) are adjacent (i.e., $\{(x_1, x_2), (y_1, y_2)\} \in E$) if and only if one of the following hold:

(i) $x_1 = y_1$ and $\{x_2, y_2\} \in E_2$;

(ii) $x_2 = y_2$ and $\{x_1, y_1\} \in E_1$;

(iii) $\{x_1, y_1\} \in E_1$ and $\{x_2, y_2\} \in E_2$.

Before moving on, let's construct the strong product of two small graphs.

Example 7.3. Let $G_1 = (\{a, b\}, \{\{a, b\}\})$ and let $G_2 = (\{1, 2, 3, 4\}, \{\{1, 2\}, \{2, 3\}, \{2, 4\})$. In Figure 7.3 we present $G_1 \boxtimes G_2$ at the top, with G_1 and G_2 given for reference. We have chosen to give $G_1 \boxtimes G_2$ a 3-dimensional representation. We encourage the reader to verify the correctness of $G_1 \boxtimes G_2$ using Definition 7.2.

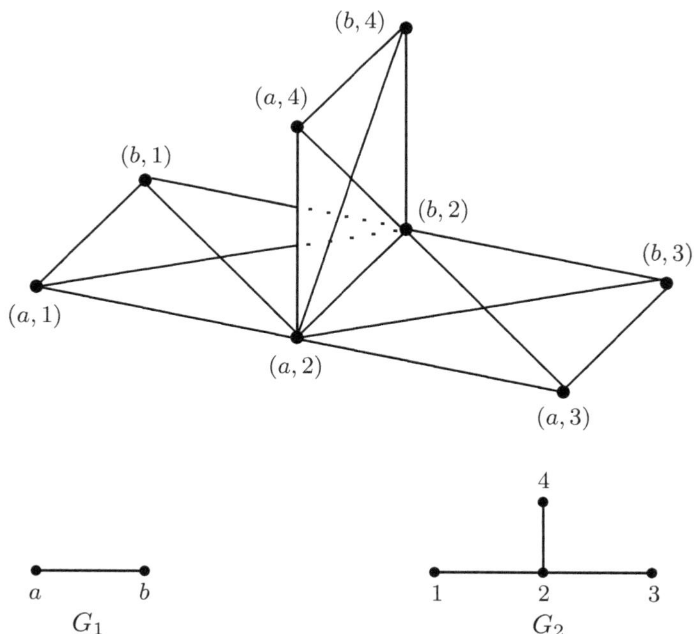

FIGURE 7.3
The strong product $G_1 \boxtimes G_2$ at top; G_1 and G_2 at bottom

By construction, the following lemma holds. We leave its proof to the reader as Exercise 7.2.

Lemma 7.4. *Let G be a confusion graph for alphabet A. Then*

$$\boxtimes_{i=1}^{k} G = \underbrace{G \boxtimes G \boxtimes \cdots \boxtimes G}_{k \text{ times}}$$

is the confusion graph for strings of length k (over A).

Let's put things together by finding the confusion graph for strings of length 2 and determining the largest independent set in $G \boxtimes G$.

Example 7.5. Consider the confusion graph of $\{a, b, c, d\}$ (say, confused by dint of proximity in the standard alphabet) given in Figure 7.4.

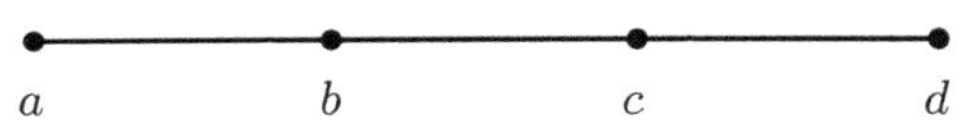

FIGURE 7.4
Confusion graph for Example 7.5

We construct the strong product $G \boxtimes G$ in Figure 7.5.

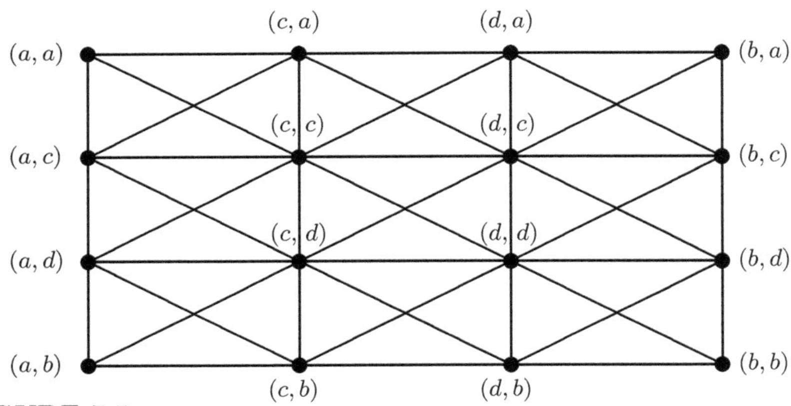

FIGURE 7.5
The strong product of the confusion graph in Figure 7.4

To transmit a string of length 2 over $\{a, b, c, d\}$ unambiguously, we want to find an independent set of vertices in the strong product graph in Figure 7.5. Searching this graph, we find that $\{(a, a), (b, b), (c, d), (d, a)\}$ is such a set. An easy pigeonhole argument shows we can do no better. So, we could now send the message “cd” through a noisy channel by sending (c, d) then (d, a) and receive the message correctly.

The confusion graph in Example 7.5 is extremely small and simple, so determining an independent set of vertices of maximal size is not difficult. What can we say for general confusion graphs? Hedrlín [104] provided a partial answer with the next theorem.

Theorem 7.6. *Let G be a graph with largest independent set of size k and let H be a graph with largest independent set of size ℓ. Then the largest independent set of $G \boxtimes H$ has size less than the Ramsey number $R(k+1, \ell+1)$.*

Proof. Suppose, for a contradiction, that $G \boxtimes H$ has an independent set of vertices I of size $N \geq R(k+1, \ell+1)$. Let $(v_1, w_1) \neq (v_2, w_2)$ be vertices in I. Since there is no edge between these two vertices, at least one of the following must hold

(i) $v_1 \neq v_2$ and $\{v_1, v_2\}$ is not an edge in G;

(ii) $w_1 \neq w_2$ and $\{w_1, w_2\}$ is not an edge in H.

Note that these are the only two options, but that both could be satisfied.

Consider the complete graph on I with a red edge between vertices that satisfy (i) and a blue edge otherwise (i.e., only (ii) is satisfied). By Ramsey's Theorem, either a red K_{k+1} or a blue $K_{\ell+1}$ subgraph exists.

If we have a red K_{k+1} subgraph, call it R, then consider

$$T = \{v \in V(G) : (v, w) \in V(R) \text{ for some } w \in H\}.$$

If $v_1, v_2 \in T$, then by the defined coloring $v_1 \neq v_2$ and $\{v_1, v_2\}$ is not an edge in G. Hence, T is an independent set of vertices of G with size $k+1$, contradicting the given size of the largest independent set of vertices of G.

If we have a blue $K_{\ell+1}$, call it B, then by using the same reasoning we see that

$$\{w \in V(H) : (v, w) \in V(B) \text{ for some } v \in G\}$$

contains an independent set of vertices of H of size $\ell+1$, also a contradiction.

Hence, we conclude that $N < R(k+1, \ell+1)$. □

Remark. Using standard graph-theoretic notation (see Section 3.2.1), Theorem 7.6 can be stated as: Let G and H be graphs. Then

$$\alpha(G \boxtimes H) < R(\alpha(G)+1, \alpha(H)+1).$$

Much more work in this subject area has been done. We refer the reader to [4, 168, 215] for further information.

7.3 Data Mining

Our next, and last, topic is a cautionary tale for data miners and those making inferences from large data sets. Data mining deals with ways of analyzing large data sets to find patterns to gain insight. As we know, Ramsey theory concerns itself with the emergence of patterns in large-enough structures. Hence, in order to gain meaningful insight from data mining, we must account for patterns that emerge due solely to the size of the data set.

This issue has only recently begun to be addressed. As a quick example, any DNA sequence is (in general) a very long sequence over the alphabet $\{A, C, G, T\}$. This is just a 4-coloring of $[1, n]$ with n very large. By van der Waerden's Theorem, we know that there will be (many) long monochromatic arithmetic progressions. These patterns in the DNA sequence do not necessarily have any meaning what-so-ever. They will occur in any DNA sequence of sufficient length.

If we consider, instead of the alphabet $\{A, C, G, T\}$, the standard English alphabet and replace the DNA sequence with any large book, the same phenomenon will occur. Brendan McKay was at the forefront of debunking the thesis of the popular book *The Bible Code*, in which the author finds "hidden messages" in the Bible. These hidden messages are referred to as *equally spaced letters*. These are just arithmetic progressions.

To be a little more concrete, let's consider finding the word "math" as an arithmetic progression in the Bible. The Bible (in English) has around 2.75 million letters in it. So, if we consider $n = 2.75$ million and let $S \subseteq [1, n]$ be the set of positions containing one of the letters $a, h, m,$ or t, we can assume that $|S| \approx \frac{4}{26}n$ (assuming each letter is equally likely, which is clearly not the case, but for our purposes here this should be fine). Since n is quite large, by Szemerédi's Theorem, S contains long arithmetic progressions.

Now, in Ramsey theory it is hard to show a *particular* monochromatic structure (e.g., red arithmetic progression) is guaranteed to exist; rather, we show one of a limited set of monochromatic structures exist (e.g., red *or* blue arithmetic progression). Hence, if we want to show that "math" exists, we should turn to probability since we can create an arbitrarily long text where "math" does not exist in arithmetic progression.

Let's assume that our value of n is large enough for Szemerédi's Theorem to afford us a 42-term arithmetic progression. So, we have a 42-term arithmetic progression of positions each with one of the letters $a, h, m,$ and t. From Exercise 6.2 we find that the expected number of occurrences of the word "math" is more than 1. Hence, observing "math" in arithmetic progression is not unusual. And there is nothing inherent in the words of the Bible; any sufficiently long text would work. For more statistical analysis involving the Bible code, the reader is referred to [143].

The confounding Ramsey theory variable in data mining seems to have only been addressed in the data mining literature recently by Calude and Longo [35] in 2017 and Pawliuk and Waddell [158] in 2019. The issue being addressed in both papers concerns spurious correlations, i.e., relationships in which variables are correlated but not in any causal way. More mathematically, in [35] we find the following definition.

Definition 7.7 (Spurious correlation). We say that a correlation is *spurious* if it appears in a randomly generated database.

Those authors also note that "the more data, the more arbitrary, meaningless and useless (for future action) correlations will be found in them." Using this framework, in [158] we are presented with a method for addressing the Ramsey paradigm. After some preliminaries, we present a sketch of one of their examples and refer the reader to [158] for more details.

In Exercise 3.4, you are asked to show a result due to Goodman [81] stating that every 2-coloring of the edges of K_n admits at least $\frac{n^3}{24}(1+o(1))$ monochromatic triangles. More accurately, Goodman proved the following theorem.

Theorem 7.8. *For any 2-coloring of the edges of K_n, there exist at least $M(n)$ monochromatic triangles, where*

$$M(n) = \begin{cases} \dfrac{m(m-1)(m-2)}{3} & \text{for } n = 2m; \\[2ex] \dfrac{2m(m-1)(4m+1)}{3} & \text{for } n = 4m+1; \\[2ex] \dfrac{2m(m+1)(4m-1)}{3} & \text{for } n = 4m+3. \end{cases}$$

Using this result, Pawliuk and Waddell [158] contend that a graph G has potentially meaningful correlation only if

$$\frac{\text{number of monochromatic triangles in } G}{M(n)} > 1.$$

We will apply this to a problem with an obvious graph model. For an example with a less obvious model, see [158].

The statistical idea being employed is to use Goodman's result to test the hypothesis that the number of monochromatic triangles behave as in a random graph versus the alternative hypothesis that they do not (in which case we could have meaningful correlations). Consider the complete graph on n vertices and color an edge red with probability p. Let T_r be the random variable equal to the number of triangles with all red edges, and let T_b be the random variable equal to the number of triangles with all blue edges. Then

$$\mathbb{E}(T_r) = \binom{n}{3}p^3 \quad \text{and} \quad \mathbb{E}(T_b) = \binom{n}{3}(1-p)^3. \tag{7.1}$$

Consider two groups of people A and B for which people in each group tend to vote alike (e.g., congressional Democrats and congressional Republicans). One measure of how consistent their respective voting practices are can be obtained by creating a complete graph on vertices $A \cup B$ with weighted edges, where the weights quantify differences in voting records.

For each member of $A \cup B$, suppose we have a sequence of votes of length n. Such a sequence can be represented as a binary string of length n, and we can quantify differences between two binary strings by the number of positions of disagreement (i.e., the Hamming distance). We place this quantity on the edge connecting the two vertices and call it an *edge-weight*. The resulting graph is called a *weighted graph*. The question now becomes: How do we turn this into a 2-coloring of edges? We expect that the subgraph on A will tend to have low edge-weights, and similarly for the subgraph on B, while all other edge-weights will tend to be higher.

In [158], the following concept is used to create 2-colored complete graphs.

Definition 7.9 (Threshold graph). Consider a weighted graph with non-negative edge-weights on vertices $v_1, v_2, \ldots, v_n$. The 2-colored *threshold graph* $G(t)$ is defined by coloring the edge between v_i and v_j red if and only if the weight of the edge between v_i and v_j is at most t (and blue otherwise).

By varying t in Definition 7.9 we obtain a sequence of 2-colored graphs that we can use for comparison purposes. For $t = 0$ we obtain a blue K_n if no two voting records are the same; for t equal to the largest weight we obtain a red K_n. What happens between these two values, and how it relates to the minimal number of guaranteed monochromatic triangles, is what we are interested in.

One quick measure is to find the minimal number of monochromatic triangles for $0 \leq t \leq W$, where W is the maximal weight. We can then compare this against $M(n)$, the minimum guaranteed by Theorem 7.8. This would mean that we have some weight threshold (voting disagreement) where many 3-person cliques have similar voting records and many 3-person cliques have mutually dissimilar voting records. Intuitively, in our congressional set-up we have many members of A voting similarly, many member of B voting similarly, and probably not many 3 person groups with a high degree of disagreement between each pair, since at least 2 people would both be in A or both be in B. To hash out what is happening a little finer, we can do the same process for the subgraph on vertices in A, and the subgraph on vertices in B. If we have significantly more monochromatic triangles than $M(n)$ we could conclude what we suspect: that members of A tend to vote alike and members of B tend to vote alike.

The issue with this approach is that we are comparing a single number (observed minimum) to another single number ($M(n)$) and we have no idea if the difference is actually statistically significant. Going back to Definition 7.7 we should measure how a random graph compares to $M(n)$ and then compare the resulting differences to our sequence of threshold graphs.

Let $R_n(p)$ be the 2-colored complete graph with each edge colored red with probability p (and blue with probability $1-p$). Assume that our weighted graphs have edge-weights between 0 and W, inclusive. Our goal is to determine an appropriate value of p for each threshold graph $G(t)$ for $t = 0, 1, \ldots, W$.

One way is to consider each voting record, i.e., binary string of length n, to be equally likely and construct the probability mass function for the possible Hamming distances. Another way is to let the possible edge-weights be equally likely, i.e., the probability of a particular edge-weight is $\frac{1}{W+1}$. There are valid arguments for both depending on the situation. The authors of [158] chose the latter, and we will follow that. As such, we will consider the random graphs $R_n\left(\frac{t+1}{W+1}\right)$, for $t = 0, 1, \ldots, W$, for our analysis.

Using the equations in (7.1) along with Theorem 7.8 we have our distribution to use for comparison:

$$X(t) = \binom{n}{3}\left(\left(\frac{t+1}{W+1}\right)^3 + \left(\frac{W-t}{W+1}\right)^3\right) - M(n), \quad t = 0, 1, \ldots, W.$$

We now enumerate the number of monochromatic triangles in each threshold graph $G(t)$, $t = 0, 1, \ldots, W$. Letting $Y(t)$ be $M(n)$ less than this count, we want to compare $Y(t)$ with $X(t)$ over all values of t. Hence, we use a χ^2 goodness-of-fit test. Thus, our test-statistic for testing whether or not $Y(t)$ behaves as our random graphs do (in terms of number of monochromatic triangles above the guaranteed minimum) is

$$\chi^2_W = \sum_{t=0}^{W} \frac{(Y(t) - X(t))^2}{X(t)}.$$

As noted before, this test-statistic can be used on just the subgraphs on vertices A and on vertices B. Perhaps the more interesting question is whether or not we can determine which group's members, A or B, vote more similarly than the other group's members. This would be a comparison of two χ^2 test-statistics.

Suppose that the weighted subgraph on vertices A, respectively B, has maximal weight W_A, respectively, W_B. Let $\chi^2_{W_A}$ and $\chi^2_{W_B}$ be the respective test-statistics. To compare these we use the F test-statistic:

$$\frac{W_B}{W_A} \cdot \frac{\chi^2_{W_A}}{\chi^2_{W_B}} \sim F_{W_A, W_B}.$$

Using threshold graphs along with Goodman's monochromatic triangles result (Theorem 7.8) we have developed hypothesis tests to aid in determining whether or not correlations are spurious. Perhaps a better development doesn't count triangles but larger monochromatic complete subgraphs. The problematic issue with this is that we have no result similar to Theorem 7.8 for the next case K_4, let alone general K_k.

Recently, experimental evidence has been provided in [170] that strongly suggests that the distribution of the number of monochromatic K_k over all graphs follows a Delaporte distribution very closely. Perhaps this could be used to decide whether or not correlations observed in large data sets are spurious.

Here's where the story ends.

–Harriet Wheeler & David Gavurin

7.4 Exercises

7.1 It is known that $x^4 + y^4 = z^2$ has no solution in the integers. Show, using Proof 2 of Theorem 7.1 as a guide, that for all sufficiently large primes p, we can find a solution to $x^4 + y^4 \equiv z^2 \pmod{p}$ with $xyz \not\equiv 0 \pmod{p}$.

7.2 Prove Lemma 7.4; that is; show that if G is a confusion graph for alphabet A, then $\boxtimes_{i=1}^{k} G$ is the confusion graph for strings of length k over A.

7.3 Let G be a simple path on n vertices. Describe $G \boxtimes G$ and determine its largest independent set.

7.4 Obtain congressional voting data from
`http://archive.ics.uci.edu/ml/datasets.`
After deleting all members with ?(s) in their voting record, perform both the χ^2 and F hypothesis tests (at a significance level of .05) as described in Section 7.3. You'll most probably want to use a computer for this. Interpret your results.

7.5 Consider the space of 2-colorings of the edges of K_{14}. Let X be the random variable giving the number of monochromatic K_4 subgraphs. Perform a random sampling of the space determining X for each sample. Take a sample size of 10^6. Create the resulting histogram of X.

Consider the random variable $D = D(\lambda, \alpha, \beta)$ with probability mass function

$$\mathbb{P}(D = j) = \sum_{i=0}^{j} \frac{\Gamma(\alpha + i)}{\Gamma(\alpha) i!} \left(\frac{\beta}{1+\beta}\right)^i \left(\frac{1}{1+\beta}\right)^\alpha \frac{\lambda^{j-i} e^{-\lambda}}{(j-i)!},$$

where $\Gamma(z)$ is the standard $\Gamma(z)$ function (where $\Gamma(n) = (n-1)!$ for positive integers). Such a random variable is called a *Delaporte random variable*. Graph $D(16.75, 7.89, 1.84)$ and compare with your histogram of X. See [170] for more information.

7.6 Let p be a prime and r-color $\mathbb{Z}^+$ as follows. For $i \in \mathbb{Z}^+$, write $i = p^k \ell$ with $p \nmid \ell$. Color i by k modulo r. Applying Hindman's Theorem, what property does the guaranteed monochromatic set have under this coloring?

7.7 Appealing only to Ramsey's Theorem, prove that for any r-coloring of $\mathbb{Z}^+$ there exists a sequence of positive integers $s_1, s_2, \ldots, s_n$, not necessarily distinct, such that for any $i, j \in [1, n]$ with $i \leq j$ all sums

$$s_i + s_{i+1} + \cdots + s_j$$

are the same color.

7.8 Use the infinite Ramsey Theorem to show that any sequence of real numbers admits an infinite monotonic subsequence.

7.9 Let X be a finite topological space. Show that there exists a finite topological space Y such that every r-coloring of Y admits a monochromatic subspace X' that is homeomorphic to X.

Hint:

Let $Y = X^n$ with the product topology and n large enough to use the Hales-Jewett Theorem.

7.10 Consider the Preface of this book (on page xi). Show that it contains the words "buy" and "this" as equally spaced letter messages (i.e., in arithmetic progression). The rules of this game are that the words are allowed to appear in reverse order (yub and siht) and we disregard non-letters (i.e., spaces, punctuation, etc.). Also, do not count the actual word "this" (that's cheating). Deduce that, upon reading this book's Preface, you had no choice but to buy this book.

Bibliography

[1] H. Abbott and D. Hanson, A problem of Schur and its generalizations, *Acta Arithmetica* **20** (1972), 175-187.

[2] M. Ajtai, J. Komlós, M. Simonovits, and E. Szemerédi, Exact solution of the Erdős-Sós conjecture, unpublished; see `https://imada.sdu.dk/Research/GT2015/Talks/Slides/simonovits.pdf`.

[3] B. Alexeev and J. Tsimerman, Equations resolving a conjecture of Rado on partition regularity, *J. Combinatorial Theory, Series A* **117** (2010), 1008-1010.

[4] N. Alon and A. Orlitsky, Repeated communication and Ramsey graphs, *IEEE Transactions Information Theory* **41** (1995), 1276-1289.

[5] N. Alon and J. Spencer, *The Probabilistic Method*, fourth edition, Wiley Series in Discrete Mathematics and Optimization, Wiley, New Jersey, 2016.

[6] K. Appel and W. Haken, The solution of the four-color-map problem, *Scientific American* **237** (1977), 108-121.

[7] A. Appel and W. Haken, Every planar map is four colorable. Part I. Discharging, *Illinois J. Math.* **21** (1977), 429-490.

[8] K. Appel, W. Haken, and J. Koch, Every planar map is four colorable. Part II. Reducibility, *Illinois J. Math.* **21** (1977), 491-567.

[9] V. Arnautov, Nondiscrete topologizability of countable rings, *Soviet Math. Doklady* **11** (1970), 423-426.

[10] L. Baumert, Sum-free sets, *Jet Propulsion Lab. Res. Summary, No. 36-10* **1** (1961), 16-18.

[11] J. Baumgartner, A short proof of Hindman's theorem, *J. Combinatorial Theory, Series A* **17** (1974), 384-386.

[12] J. Beck, An algorithmic approach to the Lovász local lemma, I, *Random Structures Algorithms* **2** (1991), 343-365.

[13] J. Beck, W. Pegden, and S. Vijay, The Hales-Jewett number is exponential – game-theoretic consequences, in *Analytic Number Theory: Essays in Honour of Klaus Roth*, 22-37, Cambridge University Press, 2009.

[14] F. Behrend, On sequences of integers containing no arithmetic progression, *Časopis Pro Pěstováni Matematiky a Fysiky* **67** (1938), 235-239.

[15] F. Behrend, On sets of integers which contain no three terms in arithmetic progression, *Proceedings National Academy Science* **32** (1946), 331-332.

[16] A. Belov and S. Okhitin, On a combinatorial problem, *Russian Math. Surveys* **48** (1993), 170.

[17] V. Bergelson, Ergodic theory of $\mathbb{Z}^d$-action, *London Math. Society Lecture Note Series* **228** (1996), 1-61.

[18] V. Bergelson, A density statement generalizing Schur's theorem, *J. Combinatorial Theory, Series A* **43** (1986), 338-343.

[19] V. Bergelson, N. Hindman, and R. McCutcheon, Notions of size and combinatorial properties of quotient sets in semigroups, *Topology Proceedings* **23** (1998), 23-60.

[20] E. Berlekamp, A construction for partitions which avoid long arithmetic progressions, *Canadian Math. Bulletin* **11** (1968), 409-414.

[21] A. Beutelspacher and W. Brestovansky, Generalized Schur numbers, in *Combinatorial Theory* (D. Jungnickel and K. Vedder, editors), 30-38, Lecture Notes in Mathematics **969**, Springer, Berlin, 1982.

[22] G. Birkhoff, *Dynamical Systems*, American Math. Society, Colloquium Pubs. **9**, Providence, 1927.

[23] T. Blankenship, J. Cummings, and V. Taranchuk, A new lower bound for van der Waerden numbers, *European J. Combinatorics* **69** (2018), 163-168.

[24] T. Bloom, A quantitative improvement for Roth's theorem on arithmetic progressions, *J. London Math. Society* **93** (2016), 643-663.

[25] T. Bloom and O. Sisack, Breaking the logarithmic barrier in Roth's theorem on arithmetic progressions, preprint available at `arXiv:2007:03528`.

[26] T. Bohman and P. Keevash, The early evolution of the H-free process, *Inventiones Math.* **181** (2010), 291-336.

[27] M. Bóna, A Euclidean Ramsey theorem, *Discrete Math.* **122** (1993), 349-352.

[28] M. Bóna and G. Tóth, A Ramsey-type problem on right-angled triangles in space, *Discrete Math.* **150** (1996), 61-67.

[29] T. Brown, An interesting combinatorial method in the theory of locally finite semigroups, *Pacific J. Math.* **36** (1971), 285-289.

[30] T. Brown and V. Rödl, Monochromatic solutions to equations with unit fractions, *Bulletin Australian Math. Society* **43** (1991), 387-392.

[31] S. Burr, A survey of noncomplete Ramsey theory for graphs, *Annals New York Academy Science* **328** (1979), 58-75.

[32] S. Burr, Generalized Ramsey theory for graphs – a survey, *Lecture Notes Math.* **406** (1974), 52-75.

[33] S. Burr and S. Loo, On Rado numbers I, unpublished.

[34] S. Butler, K. Costello, and R. Graham, Finding patterns avoiding many monochromatic constellations, *Experimental Math.* **19** (2010), 399-411.

[35] C. Calude and G. Longo, The deluge of spurious correlations in big data, *Found Science* **22** (2017), 595-612.

[36] C. Carathéodory, Über den wiederkehrsatz von Poincaré, *Sitzungsberichte Preussischen Akademie Wissenschaften* **32** (1919), 580-584.

[37] H. Cartan, Théorie des filters, *Comptes Rendus l'Académie Science Paris* **205** (1937), 595-598.

[38] E. Čech, On bicompact spaces, *Annals Math.* **38** (1937), 823-844.

[39] R. Chacon, Weakly mixing transformations which are not strongly mixing, *Proceedings American Math. Society* **22** (1969), 559-562.

[40] S. Chow, S. Lindqvist, and S. Prendiville, Rado's criterion over squares and higher powers, *J. European Math. Society*, to appear; preprint available at `arXiv:1806:05002`.

[41] N. Chudakov, On the Goldbach problem, *Doklady Akademii Nauk SSSR* **17** (1937), 335-338.

[42] N. Chudakov, On the density of the set of even numbers which are not representable as a sum of two odd primes, *Izvestiya Rossiiskoi Akademii Nauk SSSR* **2** (1938), 25-40.

[43] F. Chung, On the Ramsey numbers $N(3,3,\ldots,3;2)$, *Discrete Math.* **5** (1973), 317-321.

[44] A. Church, An unsolvable problem of elementary number theory, *American J. Math.* **58** (1936), 345-363.

[45] V. Chvátal, Tree-complete graph Ramsey numbers, *J. Graph Theory* **1** (1977), 93.

[46] V. Chvátal and F. Harary, Generalized Ramsey theory for graphs, III. Small off-diagonal numbers, *Pacific J. Math.* **41** (1972), 335-345.

[47] M. Codish, M. Frank, A. Itzhakov, and A. Miller, Computing the Ramsey number $R(4,3,3)$ using abstraction and symmetry breaking, *Constraints* **21** (2016), 375-393.

[48] W. Comfort, Ultrafilters: Some old and some new results, *Bulletin American Math. Society* **83** (1977), 417-455.

[49] D. Conlon, A new upper bound for diagonal Ramsey numbers, *Annals Math., Second Series* **170** (2009), 941-960.

[50] D. Conlon and A. Ferber, Lower bounds for multicolor Ramsey numbers, preprint available at `arXiv:2009.10458`.

[51] D. Conlon, J. Fox, and B. Sudakov, Hypergraph Ramsey numbers, *J. American Math. Society* **23** (2010), 247-266.

[52] E. Croot III, On a coloring conjecture about unit fractions, *Annals Math.* **157** (2003), 545-556.

[53] P. Csikvári, A. Sárközy, and K. Gyarmai, Density and Ramsey type results on algebraic equations with restricted solution sets, *Combinatorica* **32** (2012), 425-449.

[54] N. de Bruijn and P. Erdős, A color problem for infinite graphs and a problem in the theory of relations, *Indagationes Math.* **13** (1951), 369-373.

[55] A. de Grey, The chromatic number of the plane is at least 5, *Geombinatorics* **28** (2018), 5-18.

[56] W. Deuber, Partitionen und lineare gleichungssysteme, *Math. Zeitschrift* **133** (1973), 109-123.

[57] L. Dickson, Lower limit for the number of sets of solutions of $x^e + y^e + z^e \equiv 0 \pmod{p}$, *J. fur Reine und Angewandte Math.* **135** (1901), 181-188.

[58] M. Di Nasso and L. Luperi Baglini, Ramsey properties of nonlinear Diophantine equations, *Advances in Math.* **324** (2018), 84-117.

[59] G. Dirac, *On the Colouring of Graphs: Combinatorial Topology of Linear Complexes*, Ph.D. thesis, University of London, 1952.

[60] A. Doss, D. Saracino, and D. Vestal Jr., An excursion into nonlinear Ramsey theory, *Graphs and Combinatorics* **29** (2013), 407-415.

[61] M. Elkin, An improved construction of progression-free sets, *Israel J. Math.* **184** (2011), 93-128.

[62] R. Ellis, Distal transformation groups, *Pacific J. Math.* **8** (1958), 401-405.

[63] P. Erdős, Some remarks on the theory of graphs, *Bulletin American Math. Society* **53** (1947), 292-294.

[64] P. Erdős, R. Graham, P. Montgomery, B. Rothschild, J. Spencer, and E. Straus, Euclidean Ramsey theorems I, *J. Combinatorial Theory, Series A* **14** (1973), 341-363.

[65] P. Erdős, R. Graham, P. Montgomery, B. Rothschild, J. Spencer, and E. Straus, Euclidean Ramsey theorems III, in *Infinite and Finite Sets II* (A. Hajnal, R. Rado, and V. Sós, editors), 559-583, Colloquia Math. Societatis János Bolyai, Amsterdam, 1975.

[66] P. Erdős, C. Ko, and R. Rado, Intersection theorems for systems of finite sets, *Quarterly J. Math.* **12** (1961), 313-320.

[67] P. Erdős and L. Lovász, Problems and results on 3-chromatic hypergraphs and some related questions, in *Infinite and Finite Sets II* (A. Hajnal, R. Rado, and V. Sós, editors), 609-627, Colloquia Math. Societatis János Bolyai, Amsterdam, 1975.

[68] P. Erdős and R. Rado, Combinatorial theorems on classifications of subsets of a given set, *Proceedings London Math. Society* **3** (1952), 417-439.

[69] P. Erdős and J. Spencer, Monochromatic sumsets, *J. Combinatorial Theory, Series A* **50** (1989), 162-163.

[70] P. Erdős and G. Szekeres, A combinatorial problem in geometry, *Compositio Mathematica* **2** (1935), 463-470.

[71] P. Erdős and P. Turán, On some sequences of numbers, *J. London Math. Society* **11** (1936), 261-264.

[72] T. Estermann, Proof that every large integer is the sum of two primes, *Proceedings London Math. Society* (2) **42** (1937), 501-516.

[73] G. Exoo, D. Ismailescu, and M. Lim, On the chromatic number of $\mathbb{R}^4$, *Discrete Computational Geometry* **52** (2014), 416-423.

[74] R Faudree and R. Schelp, All Ramsey numbers for cycles in graphs, *Discrete Math.* **8** (1974), 313-329.

[75] J. Fox and D. Kleitman, On Rado's boundedness conjecture, *J. Combinatorial Theory, Series A* **113** (2006), 84-100.

[76] H. Furstenberg, Ergodic behavior of diagonal measures and a theorem of Szemerédi on arithmetic progressions, *J. d'Analyse Math.* **31** (1977), 204-256.

[77] H. Furstenberg and Y. Katznelson, A density version of the Hales-Jewett theorem, *J. d'Analyse Math.* **57** (1991), 64-119.

[78] H. Furstenberg and B. Weiss, Topological dynamics and combinatorial number theory, *J. d'Analyse Math.* **34** (1978), 61-68.

[79] F. Galvin and K. Prikry, Borel sets and Ramsey's theorem, *J. Symbolic Logic* **38** (1973), 193-198.

[80] L. Gerencsér and A. Gyárfás, On Ramsey-type problems, *Annales Universitatis Scientiarum Budapestinensis, Eötvös Sect. Math.* **10** (1967), 167-170.

[81] A. Goodman, On sets of acquaintances and strangers at any party, *American Math. Monthly* **66** (1959), 778-783.

[82] W. T. Gowers, A new proof of Szemerédi's theorem, *Geometric Functional Analysis* **11** (2001), 465-588.

[83] R. Graham and S. Butler, *Rudiments of Ramsey Theory*, second edition, American Math. Society, CBMS **123**, 2015.

[84] R. Graham, B. Rothschild, and J. Spencer, *Ramsey Theory*, second edition, Wiley-Interscience, Discrete Math. and Optimization, New York, 1990.

[85] R. Graham and E. Tressler, Open problems in Euclidean Ramsey theory, in *Ramsey Theory: Yesterday, Today, and Tomorrow* (A. Soifer, editor), 115-120, Birkhäuser/Springer, New York, 2011.

[86] A. Granville, An introduction to additive combinatorics, in *Additive Combinatorics* (A. Granville, M. Nathanson, and J. Solymosi, editors), 1-27, American Math. Society, CRM Proceedings and Lecture Notes **43**, Providence, Rhode Island, 2007.

[87] B. Green, Book review: Additive Combinatorics, *Bulletin Amer. Math. Society* **46** (2009), 489-497.

[88] B. Green and S. Lindqvist, Monochromatic solutions to $x + y = z^2$, *Canadian J. Math.* **71** (2019), 579-605.

[89] B. Green and T. Sanders, Monochromatic sums and products, *Discrete Analysis J.* (2016), #5.

[90] B. Green and T. Tao, The primes contain arbitrarily long arithmetic progressions, *Annals Math.* **167** (2008), 481-547.

[91] B. Green and J. Wolf, A note on Elkin's improvement of Behrend's construction, in *Additive Number Theory* (D. Chudnovsky and G. Chudnovsky, editors), 141-144, Springer, New York, 2010.

[92] E. Grosswald, Arithmetic progressions that consist only of primes, *J. Number Theory* **14** (1982), 9-31.

[93] D. Gunderson, On Deuber's partition theorem for (m, p, c)-sets, *Ars Combinatoria* **63** (2002), 15-31.

[94] S. Guo and Z-W. Sun, Determination of the two-color Rado number for $a_1x_1 + \cdots + a_mx_m = x_0$, *J. Combinatorial Theory, Series A* **115** (2008), 345-353.

[95] S. Gupta, R. Thulasi, and A. Tripathi, The two-colour Rado number for the equation $ax + by = (a + b)z$, *Annals Combinatorics* **19** (2015), 269-291.

[96] A. Hales and R. Jewett, Regularity and positional games, *Transactions American Math. Society* **106** (1963), 222-229.

[97] R. Hancock and A. Treglown, On solution-free sets of integers, *European J. Combinatorics* **66** (2017), 110-128.

[98] R. Hancock and A. Treglown, On solution-free sets of integers II, *Acta Arithmetica* **180** (2017), 15-33.

[99] F. Harary, Recent results on generalized Ramsey theory for graphs, in *Graph Theory and Applications* (Y. Alavi, D. Lick, and A. White, editors), 125-138, Springer, Berlin, 1972.

[100] G. Hardy and J. Littlewood, Some problems of "Partitio Numerorum" I: A new solution of Waring's problem, *Nachrichten Gesellschaft Wissenschaften Göttingen* **1920** (1920), 33-54.

[101] G. Hardy and S. Ramanujan, Asymptotic formulae in combinatory analysis, *Proceedings London Math. Society* **17** (1918), 75-115.

[102] G. Hardy and E. Wright, *An Introduction to the Theory of Numbers*, fifth edition, The Clarendon Press, Oxford University Press, New York, 1979.

[103] J. Heawood, Map-colour theorems, *Quarterly J. Math., Oxford* **24** (1890), 332-338.

[104] Z. Hedrlín, An application of the Ramsey theorem to topological products, *Bulletin l'Académie Polonaise Science* **14** (1966), 25-26.

[105] M. Heule, Schur number five, in *Proceedings of the Thirty-Second AAAI Conference on Artificial Intelligence* (S. McIlraith and K. Weinberger, editors), 6598-6606, AAAI Press, New Orleans, Louisiana, 2018.

[106] M. Heule, Computing small unit-distance graphs with chromatic number 5, *Geombinatorics* **28** (2018), 32-50.

[107] M. Heule, O. Kullmann, and V. Marek, Solving and verifying the Boolean Pythagorean triples problem via cube-and-conquer, in *Theory and Applications of Satisfiability Testing* (N. Creignou and D. Le Berre, editors), 228-245 Springer, Lecture Notes in Computer Science **9710**, Switzerland, 2016.

[108] D. Hilbert, Über die irreducibilität ganzer rationaler funktionen mit ganzzahligen coefizienten (On the irreducibility of entire rational functions with integer coefficients), *J. Reine Angewandte Math.* **110** (1892), 104-129.

[109] N. Hindman, Finite sums from sequences within cells of a partition of $\mathbb{N}$, *J. Combinatorial Theory, Series A* **17** (1974), 1-11.

[110] N. Hindman, Partitions and sums and products of integers, *Transactions American Math. Society* **247** (1979), 227-245.

[111] N. Hindman and D. Strauss, *Algebra in the Stone-Čech Compactification*, second edition, De Gruyter, Berlin, 2012.

[112] N. Hindman and E. Tressler, The first nontrivial Hales-Jewett number is four, *Ars Combinatoria* **113** (2014), 385-390.

[113] V. Jelínek, J. Kynčl, R. Stolař, and T. Valla, Monochromatic triangles in two-colored plane, *Combinatorica* **29** (2009), 699-718.

[114] P. Johnson, Two consequences of the recent discovery that $\chi(\mathbb{R}^2, 1) > 4$, *Geombinatorics* **28** (2018), 87-92.

[115] R. Juhász, Ramsey type theorems in the plane, *J. Combinatorial Theory, Series A* **27** (1979), 152-160.

[116] V. Jungić, Introduction to Ramsey Theory (lecture notes), `http://people.math.sfu.ca/~vjungic/RamseyNotes/book-1.html`).

[117] M. Katz and J. Reimann, *An Introduction to Ramsey Theory: Fast Functions, Infinity, and Metamathematics*, American Math. Society, Student Math. Library **87**, Providence, Rhode Island, 2018.

[118] A. Kempe, On the geographical problem of the four colours, *American J. Math.* **2** (1879), 193-220.

[119] A. Y. Khinchin, *Three Pearls of Number Theory*, Graylock Press, Rochester, New York, 1956 (also available by Dover, 1998).

[120] P. Komjáth, Erdős's work on infinite graphs, in *Erdős Centennial* (L. Lovász, I. Ruzsa, and V. Sós, editors), 325-345, Springer, Bolyai Society Math. Studies **25**, Berlin, Germany, 2013.

[121] P. Komjáth and V. Totik, Ultrafilters, *American Math. Monthly* **115** (2008), 33-44.

[122] D. König, *Theorie der Endlichen und Unendlichen Graphen*, Leipzig: Akademische Verlagsgesellschaft, 1936.

[123] D. König, *Theory of Finite and Infinite Graphs*, Birkhäuser, Berlin, Germany, 1986.

[124] M. Kouril, Computing the van der Waerden number $W(3,4) = 293$, *Integers* **12** (2012), #A46.

[125] M. Kouril and J. Paul, The van der Waerden number $W(2,6)$ is 1132, *Experimental Math.* **17** (2008), 53-61.

[126] J. Kozik and D. Shabanov, Improved algorithms for colorings of simple hypergraphs and applications, *J. Combinatorial Theory, Series B* **116** (2016), 312-332.

[127] N. Krylov and N. Bogoliubov, La théorie generalie de la mesure dans son application á l'étude de systámes dynamiques de la mécanique non-linéaire, *Annals Math.* **38** (1937), 65-113.

[128] C. Lambie-Hanson, Ultrafilters VIII: Chromatic compactness, `http://pointatinfinityblog.wordpress.com/2017/01/10/`.

[129] B. Landman and A. Robertson, *Ramsey Theory on the Integers*, second edition, American Math. Society, Student Math. Library **73**, Providence, Rhode Island, 2014.

[130] I. Leader, Lecture notes on Ramsey theory, University of Cambridge, `https://www.dpmms.cam.ac.uk/~par31/notes/ramsey.pdf`, 2000.

[131] K. Leeb, A full Ramsey-theorem for the Deuber-category, in *Infinite and Finite Sets II* (A. Hajnal, R. Rado, and V. Sós, editors), 1043-1049, Colloquia Math. Societatis János Bolyai, Amsterdam, 1975.

[132] H. Lefmann, A note on Ramsey numbers, *Studia Scientiarum Math. Hungarica* **22** (1987), 445-446.

[133] H. Lefmann, On partition regular systems of equations, *J. Combinatorial Theory, Series A* **58** (1991), 35-53.

[134] Y. Li and J. Shu, A lower bound for off-diagonal van der Waerden numbers, *Advances Applied Math.* **44** (2010), 243-247.

[135] S. Lindqvist, Partition regularity of generalised Fermat equations, *Combinatorica* **38** (2018), 1457-1483.

[136] A. Lott, Roth's theorem on arithmetic progressions, undergraduate thesis, 2017; available at `http://math.ucla.edu/adamlott99`.

[137] L. Lovász, *Combinatorial Problems and Exercises*, second edition, American Math. Society Chelsea, Providence, Rhode Island, 2007.

[138] L. Lu, A. Mohr, and L. Székely, Quest for negative dependency graphs, in *Recent Advances in Harmonic Analysis and Applications* (B. Bilyk, L. De Carli, A. Petukhov, A. Stokolos, and B. Wick, editors), 243-258, Springer, New York, 2012.

[139] W. Luxemburg, A remark on a paper by N. G. de Bruijn and P. Erdős, *Indagationes Math.* **24** (1962), 343-345.

[140] N. Lyall, Schur's Theorem: University of Georgia REU - Additive Combinatorics – Summer 2010, available at `http://math.uga.edu/~lyall/REU2010/schur.pdf`.

[141] K. Maran, S. Reddy, D. Sharma, and A. Tripathi, Some results on a class of mixed van der Waerden numbers, *Rocky Mountain J. Math.* **48** (2018), 885-904.

[142] A. R. D. Mathias, On a generalization of Ramsey's theorem, *Notices American Math. Society* **15** (1968), 931.

[143] B. McKay, D. Bar-Natan, M. Bar-Hillel, and G. Kalai, Solving the Bible code, *Statistical Science* **14** (1999), 150-173.

[144] B. McKay and S. Radziszowski, The first classical Ramsey number for hypergraphs is computed, in *Proceedings of the Second Annual ACM-SIAM Symposium on Discrete Algorithms*, 304-308, Society for Industrial and Applied Math., Philadelphia, Pennsylvania, 1991.

[145] A. Mohr, *Applications of the Lopsided Lovász Local Lemma Regarding Hypergraphs*, Ph.D. thesis, University of South Carolina, 2013, available at `www.austinmohr.com`.

[146] M. Molloy and B. Reed, *Graph Colouring and the Probabilistic Method*, Algorithms and Combinatorics **23**, Springer, Berlin, 2002.

[147] J. Moreira, Rado's theorem and Deuber's theorem (2014), `https://joelmoreira.wordpress.com/2014/09/23/rados-theorem-and-deubers-theorem/`.

[148] J. Moreira, Monochromatic sums and products in $\mathbb{N}$, *Annals Math., Second Series* **185** (2017), 1069-1090.

[149] L. Moser and K. Moser, Solution to problem 10, *Canadian Math. Bulletin* **4** (1961), 187-189.

[150] R. Moser and G. Tardos, A constructive proof of the general Lovász local lemma, *J. Association Computing Machinery* **57** (2010), #11.

[151] K. Myers and J. Parrish, Some nonlinear Rado numbers, *Integers* **18B** (2018), #A6.

[152] J. Nešetřil and V. Rödl, Another proof of the Folkman-Rado-Sanders theorem, *J. Combinatorial Theory, Series A* **34** (1983), 108-109.

[153] K. Numakura, On bicompact semigroups, *Math. J. Okayama University* **1** (1952), 99-108.

[154] J. Oxtoby, The Poincaré recurrence theorem, in: *Measure and Category* (J. Oxtoby), 65-69, Springer, Graduate Texts in Mathematics **2**, New York, 1971.

[155] K. O'Bryant, Sets of integers that do not contain long arithmetic progressions, *Electronic J. Combinatorics* **18** (2011), #P59.

[156] P. Pach, Monochromatic solutions to $x+y=z^2$ in the interval $[N, cN^4]$, *Bulletin London Math. Society* **50** (2018), 1113-1116.

[157] P. Parrilo, A. Robertson, and D. Saracino, On the asymptotic minimum number of monochromatic 3-term arithmetic progressions, *J. Combinatorial Theory, Series A* **115** (2008), 185-192.

[158] M. Pawliuk and M. Waddell, Using Ramsey theory to measure unavoidable spurious correlations in big data, *Axioms* **8** (2019), #29.

[159] H. Poincaré, Sur le problème des trois corps et les équations de la dynamique, *Acta Math.* **13** (1890), 1-270.

[160] D. H. J. Polymath, A new proof of the density Hales-Jewett theorem, *Annals Math.* **175** (2012), 1283-1327.

[161] J. Prömel, *Ramsey Theory for Discrete Structures*, Springer, Cham, Switzerland, 2013.

[162] R. Rado, Some recent results in combinatorial analysis, *Comptes Rendus Congrès des Mathématiciens* **2** (1936), 20-21.

[163] R. Rado, Studien zur kombinatorik, *Math. Z.* **36** (1933), 424-470. Available (as Section I) at `https://edoc.hu-berlin.de/handle/18452/795`.

[164] R. Rado, Verallgemeinerung eines satzes von van der Waerden mit anwendungen auf ein problem der Zahlentheorie, *Sonderausg. Sitzungsber. Preuss. Akad. Wiss. Phys.-Math. Klasse* **17** (1933), 1-10.

[165] S. Radziszowski, Small Ramsey numbers, *Electronic J. Combinatorics*, Survey, revision #15 (2007), #DS1.

[166] F. Ramsey, On a problem of formal logic, *Proceedings London Math. Society* **30** (1930), 264-286.

[167] F. Reisz, Stetigkeitsbegriff und abstrakte Mengenlehre, *Atti IV Congresso Internaz. Mat.* **II** (Roma, 1908), R. Accad. Lincei, Rome, 1909.

[168] F. Roberts, Applications of Ramsey theory, *Discrete Applied Math.* **9** (1984), 251-261.

[169] A. Robertson, A probabilistic threshold for monochromatic arithmetic progressions, *J. Combinatorial Theory, Series A* **137** (2016), 79-87.

[170] A. Robertson, W. Cipolli, and M. Dascălu, On the distribution of monochromatic complete subgraphs and arithmetic progressions, *Experimental Math.*, **70** (2021), 135-145.

[171] A. Robertson and D. Zeilberger, A 2-coloring of $[1, n]$ can have $n^2/22 + O(n)$ monochromatic Schur triples, but not less!, *Electronic J. Combinatorics* **5** (1998), #R19.

[172] V. Rosta, On a Ramsey type problem of J. A. Bondy and P. Erdős, I & II, *J. Combinatorial Theory, Series B* **15** (1973), 94-120.

[173] V. Rosta, Ramsey theory applications, *Electronic J. Combinatorics*, Survey (2004), #DS13.

[174] K. Roth, Sur quelques ensembles d'entiers, *Comptes Rendus l'Académie Science Paris* **234** (1952), 388-390.

[175] K. Roth, Irregularities of sequences relative to arithmetic progressions, IV, *Periodica Math. Hungarica* **2** (1972), 301-326.

[176] F. Rowley, Constructive lower bounds for Ramsey numbers from linear graphs, *Australasian J. Comb.* **68** (2017), 385-395.

[177] J. Sahasrabudhe, Exponential patterns in arithmetic Ramsey theory, *Acta Arithmetica* **182** (2018), 13-42.

[178] J. Sanders, *A Generalization of Schur's Theorem*, Ph.D. thesis, Yale University, New Haven, 1968. See also `arXiv:1712.03620`.

[179] D. Saracino, The 2-color Rado number of $x_1 + x_2 + \cdots + x_{m-1} = ax_m$, II, *Ars Combinatoria* **119** (2015), 193-210.

[180] D. Saracino, The 2-color Rado number of $x_1 + x_2 + \cdots + x_n = y_1 + y_2 + \cdots + y_k$, *Ars Combinatoria* **129** (2016), 315-321.

[181] A. Sárközy, On sums and products of residues modulo p, *Acta Arithmetica* **118** (2005), 403-409.

[182] R. Salem and D. Spencer, On sets of integers which contain no three terms in arithmetical progression, *Proceedings National Academy Science* **28** (1942), 561-563.

[183] T. Schoen, The number of monochromatic Schur triples, *European J. Combinatorics* **20** (1999), 855-866.

[184] T. Schoen, A subexponential upper bound for the van der Waerden numbers $W(3,k)$, preprint available at `arXiv:2006.02877`.

[185] T. Schoen and O. Sisask, Roth's theorem for four variables and additive structures in sums of sparse sets, *Forum Math. Sigma* **4** (2016), #e5.

[186] I. Schur, Über die kongruenz $x^m + y^m = z^m \pmod{p}$, *Jahresbericht der Deutschen Mathematiker-Vereinigung* **25** (1916), 114-117.

[187] Y. Setyawan, Combinatorial Number Theory: Results of Hilbert, Schur, Folkman, and Hindman, Master's Thesis, Simon Fraser University, Canada, 1998.

[188] L. Shader, All right triangles are Ramsey in E^2, *J. Combinatorial Theory, Series A* **20** (1976), 385-389.

[189] S. Shelah, Primitive recursive bounds for van der Waerden numbers, *J. American Math. Society* **1** (1988), 683-697.

[190] A. Sisto, Exponential triples, *Electronic J. Combinatorics* **18** (2011), #P147.

[191] A. Soifer, *The Mathematical Coloring Book*, Springer, New York, 2009.

[192] K. Soundararajan, available at `www.math.cmu.edu/~af1p/Teaching/AdditiveCombinatorics/Soundararajan.pdf`.

[193] J. Spencer, Asymptotic lower bounds for Ramsey functions, *Discrete Math.* **20** (1977), 69-76.

[194] J. Spencer, Ramsey's theorem – a new lower bound, *J. Combinatorial Theory, Series A* **18** (1975), 108-115.

[195] E. Szemerédi, On sets of integers containing no four elements in arithmetic progression, *Acta Math. Academiae Scientiarum Hungaricae* **20** (1969), 89-104.

[196] E. Szemerédi, On sets of integers containing no k elements in arithmetic progression, *Acta Arithmetica* **27** (1975), 199-245.

[197] T. Tao and V. Vu, *Additive Combinatorics*, Cambridge University Press, New York, 2009.

[198] A. Tarski, Une contribution à la théorie de la mesure, *Fundamenta Math.* **15** (1930), 42-50.

[199] A. Taylor, Bounds for the disjoint unions theorem, *J. Combinatorial Theory, Series A* **30** (1981), 339-344.

[200] T. Thanatipanonda and E. Wong, On the minimum number of monochromatic generalized Schur triples, *Electronic J. Combinatorics* **24** (2017), #2.20.

[201] G. Tóth, A Ramsey-type bound for rectangles, *J. Graph Theory* **23** (1996), 53-56.

[202] A. Turing, On computable numbers, with an application to entscheidungsproblem, *Proceedings London Math. Society* **42** (1937), 230-265. Erratum: *Proceedings London Math. Society* **43** (1938), 544-546.

[203] A. Tychonoff, Über die topologische erweiterung von Räumen, *Math. Annalen* **102** (1930), 544-561.

[204] S. Ulam, Concerning functions of sets, *Fundamenta Math.* **14** (1929), 231-233.

[205] B. L. van der Waerden, Beweis einer Baudetschen vermutung, *Nieuw Archief voor Wiskunde* **15** (1927), 212-216.

[206] P. van Hoften, Primes and arithmetic progressions, Bachelor thesis, Universiteit Utrecht, available through `https://dspace.library.uu.nl`.

[207] P. Varnavides, On certain sets of positive density, *J. London Math. Society* **34** (1959), 358-360.

[208] S. Vijay, Monochromatic progressions in random colorings, *J. Combinatorial Theory, Series A* **119** (2012), 1078-1080.

[209] M. Villarino, W. Gasarch, and K. Regan, Hilbert's proof of his irreducibility theorem, *American Math. Monthly* **125**, 513-530.

[210] I. Vinogradov, Some theorems concerning the theory of primes, *Recueil Math.* **44** (1937), 179-195.

[211] I. Vinogradov, Representation of an odd number as the sum of three primes, *Doklady Akademii Nauk SSR* **15** (1937), 291-294.

[212] A. Walfisz, Zur additiven zahlentheorie, II, *Math. Zeitschrift* **40** (1936), 592-607.

[213] Z. Wang, Furstenberg's ergodic theory proof of Szemerédi's theorem, available at `http://math.uchicago.edu/~may/REU2018/REUPapers/Wang,Zijian.pdf`.

[214] A. Wiles, Modular elliptic curves and Fermat's last theorem, *Annals Math.* **142** (1995), 443-551.

[215] H. Witsenhausen, The zero-error side-information problem and chromatic numbers, *IEEE Transactions Information Theory* **IT-22** (1976) 592-593.

[216] D. Wright, Tychonoff's theorem, *Proceedings American Math. Society* **120** (1994), 985-987.

[217] X. Xiaodong, X. Zheng, and C. Zhi, Upper bounds for Ramsey numbers $R_n(3)$ and Schur numbers (Chinese), *Math. in Economics* **19** (2002), 81-84.

[218] D. Zeilberger, Symbolic moment calculus II: Why is Ramsey theory sooooo eeeenormously hard?, *Integers* **7**(2) (2007), #A34.

Index

For Product Safety Concerns and Information please contact our EU
representative GPSR@taylorandfrancis.com
Taylor & Francis Verlag GmbH, Kaufingerstraße 24, 80331 München, Germany

www.ingramcontent.com/pod-product-compliance
Ingram Content Group UK Ltd.
Pitfield, Milton Keynes, MK11 3LW, UK
UKHW021833260926
475702UK00049B/501
* 9 7 8 1 1 3 8 3 6 4 3 3 2 *